AF279228

Imported and American Varieties of Dates (*Phoenix dactylifera*) in the United States

Based on the work of Roy W. Nixon

At the height of his career, Roy W. Nixon was the world authority on dates. Photo courtesy of Stewart Nixon.

Imported and American Varieties of Dates (*Phoenix dactylifera*) in the United States

Based on the work of Roy W. Nixon

Donald R. Hodel
Environmental Hoticulturist
University of California Cooperative Extension
Los Angeles County

Dennis V. Johnson
Consultant
Cincinnati, Ohio

University of California
Agriculture and Natural Resources
Publication 3498
2007

ORDERING

For information about ordering this publication and/or a free catalog, contact

University of California
Agriculture and Natural Resources
Communication Services
6701 San Pablo Avenue, 2nd Floor
Oakland, California 94608-1239

Telephone 1-800-994-8849
(510) 642-2431
FAX (510) 643-5470
E-mail: danrcs@ucdavis.edu
Visit the ANR Communication Services Web site at http://anrcatalog.ucdavis.edu

Publication 3498

ISBN-13: 978-1-879906-78-5
ISBN-10: 1-879906-78-3
Library of Congress Control Number: 2007927242

Cover: Trees of Amir Hajj (Donald R. Hodel). Fruit insets, left to right: Khisab in *khalel* stage, Honey in *rutab* stage, Barhee in *khalal* stage (David Karp).

 Printed in Canada on recycled paper.

This publication may be cited as: Hodel, D.R. and D.V. Johnson. 2007. Imported and American Varieties of Dates (*Phoenix dactlifera*) in the United States. UC ANR Publicationi 3498. Oakland, CA: University of California.

This publication has been peer reviewed for technical accuracy by University of California scientists and other qualified professionals. This review process was managed by the ANR Associate Editor for Pomology, Viticulture, and Subtropical Horticulture.

PREFACE

The impetus for publishing this volume arose when Richard K. Harris, curator and educational program coordinator of the Arizona State University Arboretum in Tempe, brought to our attention a copy of Roy W. Nixon's unpublished (and likely unfinished) manuscript on American varieties of dates (Nixon 1955). Judging from the manuscript's content, Nixon probably last worked on it in about 1955. Why Nixon, the leading American date palm scientist of the 20th century and then at the height of his career as the world authority on dates, did not complete or publish the manuscript is unknown. He authored many other publications about dates after 1955 and he lived until 1976. His unpublished manuscript was written and formatted in much the same manner as his monumental 1950 work, *Imported Varieties of Dates in the United States,* and likely was intended to serve as a companion to that volume.

Our original intention was simply to update Nixon's manuscript and publish it, since the information about American varieties of dates had never been disseminated and so was unavailable to growers and researchers in the date industry. As we updated the manuscript, though, we decided that it would be valuable and appropriate to incorporate information about imported varieties as well, because they account for nearly all of the commercially produced dates in the United States. The most recent authoritative account of imported varieties of dates in the United States is Nixon's 1950 treatment. Thus, we felt that new and updated information would be valuable and timely for date growers and researchers as it could reflect several significant changes that occurred in the date industry over the latter half of the 20th century.

This present work, then, includes Nixon's unpublished manuscript on American varieties of dates in its entirety and borrows liberally from his 1950 work on imported varieties of dates. We have edited and updated the information from both these accounts and added critical background information, drawing upon such disciplines as history, botany, economics, and horticulture. Our intention is to provide for the first time a comprehensive account of all of the varieties of dates and the date palm industry in the United States.

This book is divided into five parts. Part one is the introduction and part two describes the botany of the date palm and its various organs. These two sections provide background information critical to understanding the past, present, and future of the date industry in the United States and to identifying and describing date varieties. Part three describes the commercial imported varieties of dates. Part four does the same for all American date varieties. We have divided American varieties into two groups: commercial and non-commercial. An addendum that includes the first published descriptions of the fruits of three new American varieties, two of which are commercial, follows Nixon's treatment on American varieties. We define *commercial* varieties as those that are available for sale, whether in large or small quantities. Although culture is usually discussed under each variety, we have included general information about date culture and management in part five.

Appendices A through D provide a biography of Nixon and a listing of his publications on dates, grading standards for dates, a list of germplasm collections of date varieties in the United States, and an identification key to important date varieties. Non-commercial imported date varieties are treated in table format in Appendix E.

The date palm, *Phoenix dactylifera,* is indigenous to the Middle East and perhaps northern Africa. There are no naturally indigenous date palms in the New World. Thus, the term "American varieties" could be seen as misleading. American varieties of dates are those that originated as seed and seedlings and were selected or developed in the United States from varieties imported primarily from the date palm's indigenous area in the Old World. All varieties, whether American or imported, are clones and must be propagated vegetatively in order to perpetuate their desirable traits. In scholarly publishing, it would be conventional to enclose date variety names within single quotation marks, but for this book we have decided to do without those marks.

Seen on a global scale, date production in the United States is minuscule. In 2003 it was only 0.21 percent of world production (14,344 of 6,776,003 metric tons), according to FAOSTAT Agriculture (http://faostat.fao.org/).

Donald R. Hodel

Dennis V. Johnson

Acknowledgments

We thank Richard Harris for bringing Roy Nixon's 1955 study of American date varieties to our attention and providing us with a copy. Stewart Nixon provided the photograph of his father reproduced in these pages. Others providing information and assistance that helped to bring this project to fruition include Doug Adair, Lee Anderson, Jr., Edie Brito, Brian Brown, Tim Burke, Kim Corrall, David Davall, Darleen DeMason, Chris Denning, Mike Falk, David Faulkner, Ted Fish, Jr., Scott Frische, the late Walter Geissler, Judy Higgins, Dwight Hurst, Roger Jacobs, Elaine Joyal, Albert Keck, Robert Krueger, Ben Laflin, Jr., Pat Laflin, Greg Leja, Hector Lukin, Juan Lukin, Peggy Mauk, Gary Nelson, Gusmar Nuñez, Elect Philip, Harry Polk, David Ritter, M. L. Robinson, Vince Samons, Gursewak Sidhu, Juan Villalobos, Charna Walker, Mardy Walton, Glenn Wright, and Duane Young. David Karp provided many of the photographs.

The International Palm Society made a generous donation to support printing of the color plates.

Production
Design and illustrations: Celeste Marquiss, ANR Communication Services
Editing: Jim Coats, ANR Communication Services

Foreword

Date palms were first introduced to the United States as seed by the Spanish missionaries in the 18th century. Establishment of a commercial industry did not occur, however, until date palm offshoots were imported by the United States Department of Agriculture (USDA) and commercial enterprises between 1890 and 1929. Although the early work of Walter Tennyson Swingle had established the climatic criteria necessary for successful date culture and had suggested appropriate varieties based upon climatic similarities between areas in the United States and the traditional areas of date cultivation, it remained for researchers to determine appropriate production practices under conditions in the United States.

Based upon the work and recommendations of Swingle, the USDA established an experimental date garden near Mecca, California, in 1904. In 1907, the USDA acquired additional land near Indio, California, and the United States Date Station and Garden was established. From the time of its establishment until its closure in 1982, the vast majority of research involving date palms in the United States was performed by Date Station personnel. The closure of the Date Station was a tribute to the success of these investigations, as in the 1970s the date industry was considered to be "mature" and no longer in need of extensive research support from the USDA.

The maturity of the date industry was due in no small part to the work of Roy Wesley Nixon. In a career that lasted from 1923 past his official retirement nearly a half century later, Nixon was active in and made vital contributions to all areas of date research. Nixon's work established cultural practices that are still in general use today in the United States and in other countries around the world, and the six editions of his *Date Growing in the United States* are still widely referenced and quoted in the date literature. His book *Imported Varieties of Dates in the United States* is still a standard reference to the characteristics of many important commercial date varieties, and his unpublished manuscript *American Varieties of Dates* is still useful to workers in the United States.

Though they are Nixon's most important book-length contributions, the two books listed in the previous paragraph have unfortunately been unavailable to most readers for some time. The release of the book you are reading now is an updated re-issuance by Donald R. Hodel and Dennis V. Johnson, and with its release Roy Nixon's valuable work is once again readily accessible to date growers, scientists, and others with an interest in date palms. The fact that the vast majority of his work remains relevant and vital attests to its quality and to the quality of Nixon as an individual and a researcher. Over a quarter of a century after his death, Roy Nixon remains *the* towering figure in date research and an inspiration to those of us involved in the field today.

Robert R. Krueger
Curator
USDA–ARS National Clonal Germplasm Repository for Citrus and Dates
Riverside, California

CONTENTS

1

History, Trends, and Current Production of Dates in the United States

One of the world's first cultivated fruit trees, the date palm (*Phoenix dactylifera* L.) was domesticated in Mesopotamia, now Iraq, more than 5,000 years ago. It has long been one of the most important plants of arid, desert areas of northern Africa, the Middle East, and southern Asia. Appropriately called "the tree of life," it has provided food, ornament, material for shelter, fiber, and fuel, and has even been used for religious purposes in a harsh environment where relatively few other plants can grow (Dowson 1982; Nixon and Carpenter 1978; Popenoe 1973; Zaid 2002). Humans have since spread the date palm far beyond its historical range, taking it to nearly all tropical and subtropical regions of the world. In his book, Zaid (2002) provides thorough coverage of date palm history and worldwide production.

Until recently it was believed that no remaining wild forms of *Phoenix dactylifera* existed. Recent research has revealed, however, that the cultivated date palm is closely related to wild and feral date palms in the Near and Middle East and North Africa. These wild date palms are morphologically similar to domesticated cultivars, share the same climatic requirements, and can hybridize with the cultivars. Botanically, wild and feral dates are considered to be the same species (Zohary and Hopf 2000).

Wherever date palms are cultivated for fruit, a clear distinction is made between *named varieties* and *seedlings*. Named date varieties or cultivars are actually clones that must be propagated vegetatively, most commonly as offshoots, young plants that arise from the base of the mother palm. Offshoots can be carefully removed and planted to grow new palms that will perpetuate a desirable character, typically that of fruit quality. Because they are clones, all individuals of a given date variety are genetically identical and have the same palm and fruit characteristics. Recently, tissue culture (micro-culture) has been used to propagate date palms, though this practice has not gained wide popularity and acceptance in the date industry in the United States.

Seedling dates, on the other hand, arise intentionally or spontaneously from seed, are variable in their characters and not genetically identical to each other or their parents, and may or may not produce horticulturally desir-

able palms that produce high quality dates. In instances where a seedling date does grow into a horticulturally desirable palm and produces superior fruit, it can then be selected and propagated vegetatively to establish a variety and perpetuate those qualities. It is through this process of careful selection of desirable seedling date palms and subsequent vegetative propagation that all known date varieties, numbering about 3,000, have been identified, established, perpetuated, and dispersed around the world over the last 5,000 years.

DATE-GROWING REGIONS OF THE UNITED STATES

Although date palms are grown primarily for ornament in many localities across the southern United States, from Florida to California, they are grown for commercial fruit production only in California and Arizona. Commercial production areas in California include Indio, Thermal, and Mecca in the Coachella Valley (Figure 1); the Bard Valley across the Colorado River from Yuma, Arizona; and Tecopa, near Death Valley in Inyo County, the northernmost area of commercial date production in North America. Although at one point California lagged behind Arizona in date production, that situation has since reversed and California now accounts for nearly all of the dates produced in the United States.

In Arizona, date palms have been grown in the Salt River Valley, including Phoenix and Tempe, and in the Yuma area. Experimental and commercial date growing in the United States first began in Arizona and that state was an early leader in date research and production. Climatic conditions in the Salt River Valley—especially summer rain—resulted in repeated losses, however, and these quickly discouraged growers (Hilgeman 1972). Later on, urban sprawl and increasing production costs lead to the destruction of most of Arizona's remaining commercial date orchards, leading to the demise of most of the industry in that state by 1960 (Hilgeman 1972). In the Yuma area, economic conditions discouraged date plantings at first, although there has been

Figure 1. Date palms heavily laden with bunches of fruit that are protected by paper coverings are a common sight during the October harvest season in California's Coachella Valley. (Donald R. Hodel)

an increase in plantings in recent years (Nixon and Carpenter 1978). Still, date production in Arizona is small compared to that of California.

A few small, experimental plantings were made in Texas at the Texas Agricultural Experiment Station at Welasco in the lower Rio Grande Valley and at Winter Haven northeast of Loredo (Wood and Mortensen 1938). Commercial date production in Texas was never successful, though.

DATES IN THE NEW WORLD

Date palms were widespread in North Africa, the Near and Middle East, southern Asia, and a few locations in southern Europe by the time Europeans first arrived in the New World (Dowson 1982). The earliest record of a date palm in the New World is from 1513 in eastern Cuba, and the tree apparently grew from seeds the Spanish had brought (Simon 1978). The motivation for this initial introduction may have been the Christian religious use of date palm leaves in Easter processions and the concern that no source of palm leaves might otherwise be available in the New World. Fruit production could also have been a motivating factor, although all early attempts to grow dates for fruit production in Cuba and later in Central Mexico failed because of unsuitable climatic conditions.

DATES IN THE UNITED STATES

The first date palms in what would eventually become the United States appear to have grown from seed planted at Spanish missions of the Franciscan and Jesuit orders in California and Arizona, which were extensions of the mission system of northwest Mexico. Most of these missions were in coastal areas of California and the date palms grown there produced very poor quality fruit because of the relatively cool, humid weather. Date palms in a number of mission gardens, however, especially those in warmer and drier areas, produced better quality fruit, and these provided an example for and encouragement to other early settlers and farmers in what are now California and Arizona to cultivate date palms (Popenoe 1973; Toumey 1898). Many seedling dates were planted in California and Arizona after the Mexican War and the California gold rush. A leading late-19th-century book on California fruit culture included a chapter on growing dates from seed (Wickson 1889).

As these seedling date palms in the more favorable regions and climates came into bearing, they drew attention to the possibility of large-scale, commercial date production in the warm, interior areas of California and Arizona. Because the dates produced on these seedling palms were of poor or, at best, variable quality, however, prospective growers and their allies recognized that success would be more likely if they were to import named, standard, commercial varieties from traditional date-growing regions in North Africa and the Near and Middle East. A number of individuals, private companies, and government institutions in the late 19th and early 20th centuries did import standard date varieties, and in this way they established the foundation of the modern commercial date industry in the United Sates. The introduction of offshoots marginalized the modest fruit production from seedling dates, which had never been commercially significant because of the great variability in fruit characteristics and poor overall fruit quality. Seedling dates continued to be grown to a lesser extent as the standard imported varieties became established, and sometimes they even arose spontaneously in date orchards, germinated from fallen fruit. Seedling dates can still be found in and around old date gardens today.

USDA IMPORTATION OF DATE OFFSHOOTS

Offshoots of Old-World date varieties were imported to the United States as early as 1818 (Mitchell 1818) and 1876 (Swingle 1901), but none survived. In 1890 and 1891 the United States Department of Agriculture (USDA) made the first successful importation of offshoots, procuring 74 plants of about a dozen varieties from Algeria, Egypt, and Muscat (Oman) and sending them to California, Arizona, and New Mexico for trial. Although about half of the offshoots were successfully established, the effort was mostly considered a failure because the offshoots had been poorly selected, their labels had been mixed up and confused, and some were relatively useless staminate (male) palms that did not produce fruit.

Despite these early, unpromising results, interest in growing commercial date varieties persisted. J. W. Toumey of the Arizona Agricultural Experiment Station urged the USDA to make another effort to introduce offshoots of standard commercial varieties, and in 1899 and 1900 Walter Swingle, a USDA scientist, arranged for the importation of 405 offshoots from Algeria (Hilgeman 1972), mostly Deglet Noor and Rhars. These were planted at the Arizona Agricultural Experiment Station at Tempe, near Phoenix, and constituted the first importation of true-to-name varieties. Swingle (1901) also published a lengthy scientific study that gave further impetus to date palm cultivation in the United States.

Arizona established a second agricultural experiment station near Yuma in 1905. From 1900 to 1908 the USDA imported 1,180 date offshoots representing 110 varieties and planted them at the two Arizona stations. The Tempe station and its research activities are partly responsible for helping initiate the subsequent commercial development of date production in California (Popenoe 1973) and Arizona. Early date production in Arizona actually exceeded that of California.

California was also gearing up for date production at this time and shared in the testing of imported varieties. The USDA established a Date Experiment Station near Mecca, California, in 1904, but moved the headquarters to another location just west of Indio in 1907 because of rising waters in the Salton Sea. The USDA planted imported offshoots at both locations.

The USDA made several other significant importations during these early years. To procure offshoots, David Fairchild traveled to Iraq, Baluchistan (Pakistan), and Egypt in 1901 to 1902 (Fairchild 1903), while Thomas Kearney journeyed to Algeria and Tunisia in 1905 (Kearney 1906). In 1927 Swingle went to Morocco and obtained a few offshoots of Medjool (Swingle 1945), the primary date of export of that country and one that would become the second most important date variety in the United States after Deglet Noor. In 1929 Nixon traveled to Iraq and obtained several new varieties. Most of Nixon's introductions went to the Texas Agricultural Experiment Station at Welasco, and eventually some second-generation offshoots were moved to the Texas station at Winter Haven.

From 1890 to 1929 the USDA imported 1,076 lots of date offshoots containing about 20,000 individuals of 149 standard date varieties (Nixon 1950, 1971). Most of the early importations were divided between Tempe in Arizona and Mecca and Indio in California; only a few went to Texas.

COMMERCIAL IMPORTATIONS OF DATE OFFSHOOTS

The successful fruiting of USDA imports encouraged commercial growers to import standard date varieties (Table 1). One individual, Bernard G. Johnson, the first commercial grower to travel abroad to import date offshoots, played a major role in establishing commercial date growing in the United States (Nixon 1946a). A German immigrant, Johnson came to California's Coachella Valley in 1901 and almost immediately began to promote date growing among the other settlers and farmers. He encouraged the Mecca Land Company to finance his trip to Algeria in 1903 and he returned with 129 offshoots of Deglet Noor and a few other varieties that he planted on land the company provided. In 1908 he again traveled to Algeria and secured more than a dozen date varieties, some of which went to the University of Arizona Agricultural Experiment Station at Yuma. At the end of another trip in 1912 Johnson returned with about 3,000 offshoots from Algeria, nearly all of them Deglet Noor (Nixon 1946a), which he planted near Yuma. Beginning in 1912, the newly formed Coachella Valley Date Growers' Association sent Johnson to Algeria three successive years to obtain sufficient quantities of Deglet Noor for commercial plantings in the Coachella Valley. Commercial date production in California had its beginning in 1912 with the first harvesting and marketing of Deglet Noor fruit (Colley 1967).

From 1911 to 1922 about a dozen date-growing companies in California and Arizona imported date offshoots of the most promising varieties, especially Deglet Noor (Nixon 1971). H. F. Cole made the first large importation of Deglet Noor from Algeria in 1911. Paul Popenoe, assisted by his brother Wilson, made the first large importation of offshoots from Iraq in 1913. In that same year, Henry Simon traveled to Algeria for West India Gardens and arranged for the importation of 6,000 Deglet Noor offshoots (Simon 1978). From 1920 to 1922, Silas Mason, a USDA employee financed by private growers, traveled to Egypt to procure offshoots.

In total about 43,320 offshoots of standard, commercial varieties were imported from 1911 to 1922, all from Algeria, Egypt, and Iraq. The principal varieties were Deglet Noor, Halawy, Hayany, Khadrawy, Kustawy, Rhars, Saidy, and Zahidi (Nixon 1971). Although initially appearing to have considerable promise, Kustawy, Rhars, and Saidy are not among the main commercial varieties grown today.

Nixon (1950) described 160 imported date varieties in the United States. Presently only 16 are commercial varieties, and these originated in four countries: Algeria, Egypt, Iraq, and Morocco. For the purposes of this book, we define *commercial* as appearing in the trade and offered for sale, even if in small quantities. The 40 American varieties that Nixon described in his unpublished 1955 manuscript (only nine of which are commercial) originated from these imported varieties, mostly as chance seedlings but occasionally as intentional, artificial hybrids.

GROWTH OF THE DATE INDUSTRY

Launching the date industry in the United States was a major challenge. Technical problems of fruit production were readily solved, but to create a demand for dates, develop a marketing structure, and encourage investment in processing facilities required optimism, vision, and perseverance. It was the date pioneer William L. Paul who provided strong leadership to the nascent industry. Paul bought 160 acres of land on the southeast side of the Coachella Valley in 1902 with the intention of growing dates there. Other business ventures intervened, and seven years elapsed before Paul and his family settled their land and began to develop a date orchard in 1909. To succeed, Paul believed, the date industry had to produce high-quality dates graded and packed according to established standards. Moreover, dates had to be available in sufficient quantities in United States markets to compete with foreign-grown dates. Paul was certain that these objectives were achievable only through the cultivation of standard imported varieties (Shumway 1960).

Table 1. Commercial importation of date offshoots into the United States, 1903–1922

Year	Variety	Quantity	Country	Importer/company
1903	Deglet Noor, Rhars, Areshty	129	Algeria	B. G. Johnson
1904	Deglet Noor	156	Algeria	California Date Co.
1908	various	?	Algeria	B. G. Johnson
1911	Deglet Noor, Rhars	1,100	Algeria	H. F. Cole/American Date Co.
1912	Deglet Noor	1,000	Algeria	P. Popenoe/West India Gardens
1912	Deglet Noor	3,000	Algeria	H. F. Cole/American Date Co.
1913	Deglet Noor, Halawy, Khadrawy, Kustawy, Zahidi	6,000 9,000	Algeria Iraq	P. Popenoe, W. Popenoe, and H. Simon/West India Gardens
1913	Halawy, Khadrawy, Kustawy, Zahidi	2,000	Algeria	B. G. Johnson/Coachella Valley Date Growers' Association
1913	Halawy, Khadrawy, Kustawy, Zahidi	1,000	Algeria	H. F. Cole/American Date Co.
1914	Halawy, Khadrawy, Kustawy, Zahidi	5,170	Algeria	B. G. Johnson/Coachella Valley Date Growers' Association
1915	Halawy, Khadrawy, Kustawy, Zahidi	3,000	Algeria	B. G. Johnson/Coachella Valley Date Growers' Association
1920	Saidy	2,000	Egypt	S. C. Mason/Gillette-Rosenberger Date Gardens and California Date Nurseries
1921	Deglet Noor	1,000	Algeria	S. C. Mason/Gillette-Rosenberger Date Gardens and California Date Nurseries
1921	Hayany	500	Egypt	S. C. Mason/Phoenix Date Co.
1922	Saidy	7,150	Egypt	S. C. Mason/Gillette-Rosenberger Date Gardens and California Date Nurseries

Source: Nixon 1950, 1971; Simon 1978.

Paul worked diligently to unite the date growers and succeeded in forming the Coachella Valley Date Growers' Association in 1913. One of its first activities was to import date offshoots from Algeria, and these were sold to member growers at an attractive price. Displays of Deglet Noor dates at the Riverside County Fair in 1914 and at the Panama-Pacific Exposition in San Francisco the following year promoted Coachella Valley dates. In 1916, Paul made promotional tours to the Eastern United States and the Pacific Northwest to publicize California's new agricultural product. Other date growers organized an annual Date Festival and Fair to promote the date industry. The first Date Festival was held in Indio in 1917, beginning an annual tradition that continues to the present day as the Riverside County Fair and National Date Festival.

Following the example set by the Coachella Valley Date Growers' Association, other growers formed similar associations, and this in turn lead in 1924 to the founding of the Date Growers' Institute. The Institute sponsored technical meetings each year and published the presentations in an annual report. For 54 years the Date Growers' Institute provided an invaluable forum for California and Arizona date growers and processors. The set of annual reports represent the single most important source of technical information about the date industry in the United States.

A major change in the California date industry came with the emergence of the Bard Valley as a significant date growing region in the latter half of the 20th century. Up until that time, the Coachella Valley had a near monopoly on date production in California, and, while it remains the number-one date-growing region in the United States, production in the Bard Valley has significantly increased.

EARLY PROBLEMS GROWING DATES IN THE UNITED STATES

The fledgling commercial date industry in the United States faced several major problems. A serious insect pest of leaves and fruits, the *Parlatoria* or white scale (*Parlatoria blanchardii*) was inadvertently introduced on some of the first date offshoots. Native to North Africa and the Middle East (Howard et al. 2001; Oihabi 2003), it was quickly recognized as a significant pest and a threat to the date industry. In the late 1920s a campaign to eradicate *Parlatoria* scale was initiated, financed by state and federal agencies. Spraying and fumigation were unsuccessful, but defoliation and blow-torching the surface of the date palm killed the pest and its eggs. Careful and judicious inspections,

monitoring, and treatment of infested palms completely eradicated *Parlatoria* scale by 1936 (Boyden 1930; Colley 1967).

In 1937, 1949, and 1950, severe freezes struck the date-growing regions of California and Arizona. Although few palms were killed outright, most sustained damage. In some cases the damage was severe, and production dipped significantly for a year or two until the palms recovered (Nixon 1938b; Swingle 1950).

Other problems included labor shortages for harvesting fruit and caring for the palms, competition from low-priced foreign imports, variable fruit quality, and low prices during the 1930s. In an effort to aid the industry, the federal government from 1935 to 1941 provided a subsidy of 3.5 cents per pound to date growers to divert lower-grade dates into byproducts to bolster the prices for whole dates. In 1937 the United Date Growers of California, a cooperative marketing association, was formed to help the industry promote and market dates. The organization functioned well through the middle 1940s when date prices were high, but it folded when prices dropped after World War II (Geissler 1971).

A California state marketing order created in 1938 to prohibit the sale of substandard dates as whole fruit addressed the issue of poor fruit quality. That same year, the United Date Growers of California established grading standards for dates. These measures evolved into the United States Grades for Dates (USDA 1955), which are still followed today. Under these standards, dates are classified into the following grades: Fancy, Choice, Standard, Substandard, and Cull (Appendix B, Table B-1). The standards apply only to Deglet Noor, Halawy, Khadrawy, and Zahidi dates, which were the major commercial varieties when the standards were developed (Bennett 1950; Colley 1967; Peightal 1956; Postlethwaite 1938; Rock 1950; Rygg 1975). USDA inspectors monitor date-packing operations to ensure that grade standards are observed; a USDA grading manual provides guidance for inspections. The grading manual includes descriptions of Barhee and Medjool fruit, to which the formal standards do not apply, but it says nothing about grading them (USDA 1977).

Although Medjool is not subject to the formal USDA standards, growers have adopted comparable quality standards for this variety. The Bard Valley Medjool Date Growers' Association, a group of seven growers that together account for 70 percent of Medjool production in California, has established its own strict quality standards (Appendix B, Table B-2).

DATE PRODUCTION IN ARIZONA

Arizona was the early U.S. leader in date production and research, and much of the technical and scientific information that served as the basis for the establishment of the California date industry actually was developed in Arizona. Despite Arizona's early lead, though, by about 1915 California had surpassed it in date production. Even at its peak in the 1930s and 1940s, Arizona's production lagged far behind that of California.

Date production and research in Arizona can be divided into three periods (Hilgeman 1972). The first period, from 1890 to 1920, saw research and a gradual increase in interest in commercial date development.

The second period, 1920 to 1950, saw the growth of commercial development. Nearly all date-production acreage in Arizona had been planted by 1932. Unfortunately, no precise date production and acreage data appear to have been collected for Arizona during this period. Production reached its peak during World War II, with an estimated 1.5 to 2 million pounds of fruit sold each year by 300 to 400 bearing acres (Hilgeman 1972). Other accounts (Powers 1945, Tate and Hilgeman 1971) estimated there were 500 acres of bearing dates from 1920 to 1945. Johnson, Joyal, and Harris (2002), extrapolating data on how many trees it would take to produce such crops, estimated there were 1,542 bearing and non-bearing acres of dates in 1934, of which nearly one-third were named varieties and the remaining were from seedlings. Nevertheless, even at its peak Arizona date production was less than one-tenth that of California.

The third period began in 1950 when commercial date production in Arizona was rapidly declining due to damage and losses from freezes and rain, increasing production costs, competition from imports, wider availability of sugar after World War II, and expanding urbanization. Production in the Salt River Valley around Phoenix, the major date-growing region in the state, was especially hard hit and had all but ceased by 1960 (Hilgeman 1972).

Despite the demise of large-scale commercial date production in the Salt River Valley, minor, limited production still persists in the area, thanks to an opportunistic and innovative home-garden system (Johnson, Joyal, and Harris 2002). Arizona Date Gardens harvests up to 100,000 pounds of fruit annually from palms in private residential gardens. The harvested variety is Sphinx (Black Sphinx), a black, soft date that originated in the Salt River Valley. These palms are remnants of commercial date orchards that covered the region before it was subdivided into housing tracts. Developers left many of the palms for ornamental landscape use around the homes in the new subdivisions. Homeowners are paid in fruit or cash. This home-garden production system may not exist much longer because many of the palms are nearing the end of their natural, productive life and will probably be replaced with other, strictly ornamental palms or trees.

Recently, there has been an increase in planting, nearly all of it Medjool and primarily in the Yuma area (Johnson, Joyal, and Harris 2002), that has coincided with the rapid expansion in planting just across the Colorado River in California's Bard Valley. Current unofficial estimates put date production in Arizona at 500 acres of bearing and 700 acres of non-bearing palms, nearly all near Yuma.

DATE PRODUCTION IN CALIFORNIA

The American Date Company constructed the first date packing house in the Coachella Valley in 1912 and the Coachella Valley Date Growers Association built a packing house in Thermal for its members in 1917. Soon other packing facilities were built in the Coachella Valley as well as in Riverside and the Los Angeles area (Geissler 1971).

Few reliable data are available on fruit production in California before 1920. One source states that 25,000 pounds of dates were produced in 1915 (Mahmud 1958). The State of California first collected and published statistics on date production in 1920 (Table 2). That year there were 280 bearing acres of dates producing 60 tons of fruits. Year-to-year production in California varies considerably, possibly to some extent because dates have a natural tendency toward alternate bearing, in particular if the fruit bunches are not thinned. It is also possible that not all of the dates produced in a given year are harvested. Also, there is no apparent correlation over the years between fruit production and bearing acres. The figures for fruit production are probably more accurate than those for bearing acres. Production trends are best interpreted from decade to decade rather than year to year.

Average annual date production in California in the 1920s was 383 tons, steadily increasing to 2,753 tons in the 1930s, 10,773 tons in the 1940s, 19,620 tons in the 1950s, and 21,670 tons in the 1960s. Average annual production declined to 21,435 tons in the 1970s but then rose to 22,270 tons in the 1980s and 23,280 tons in the 1990s. The peak annual production, 29,000 tons, was achieved in 1985 and again in 1993. Production in 2003 was 16,400 tons on 5,300 acres. (Note: Production figures for Riverside and Imperial Counties do not always add up to the total state figure, either here in text or in Tables 2, 3, and 4. This discrepancy

Table 2. California date production and bearing acreage, 1920–2003

Year	Quantity (tons)	Bearing acres	Year	Quantity (tons)	Bearing acres
1920	60	280	1962	24,200	4,200
1921	61	440	1963	22,100	4,200
1922	105	570	1964	24,300	4,200
1923	145	620	1965	21,000	4,000
1924	220	700	1966	21,300	4,000
1925	320	740	1967	20,800	4,100
1926	520	780	1968	23,000	4,300
1927	710	800	1969	16,500	4,300
1928	820	820	1970	18,200	4,000
1929	870	880	1971	19,500	4,000
1930	1,560	940	1972	15,700	4,100
1931	1,215	1,000	1973	23,100	4,200
1932	2,160	1,200	1974	22,700	4,100
1933	2,450	1,300	1975	24,800	4,100
1934	3,160	1,400	1976	22,400	3,800
1935	3,250	1,600	1977	25,100	3,800
1936	3.970	1,900	1978	21,550	4,000
1937	3,630	2,300	1979	21,300	3,900
1938	3,530	2,600	1980	22,400	4,000
1939	2,600	3,000	1981	22,300	4,100
1940	6,200	3,200	1982	23,900	4,100
1941	5,790	3,200	1983	19,400	4,200
1942	7,740	3,400	1984	22,500	4,300
1943	10,770	3,600	1985	29,000	4,400
1944	13,190	3,700	1985	17,800	4,500
1945	6,800	3,800	1987	19,400	4,900
1946	16,720	3,900	1988	22,000	5,100
1947	10,180	3,800	1989	24,000	5,000
1948	16,240	3,900	1990	24,000	5,000
1949	14,100	4,000	1991	22,000	5,200
1950	15,060	4,200	1992	21,000	5,300
1951	18,840	4,300	1993	29,000	5,500
1952	16,500	4,600	1994	23,000	5,500
1953	17,000	4,800	1995	22,700	5,200
1954	15,400	4,700	1996	23,000	4,680
1955	25,300	4,600	1997	21,000	4,800
1956	19,200	4,600	1998	24,900	4,900
1957	23,300	4,700	1999	22,200	4,900
1958	19,600	4,100	2000	17,400	4,800
1959	26,000	4,200	2001	19,700	4,500
1960	22,100	4,200	2002	24,200	4,500
1961	21,400	4,200	2003	16,400	5,300

Source: USDA 2004.

reflects differences between the collecting and reporting methodologies used by the county and state agencies).

DATE PRODUCTION IN RIVERSIDE COUNTY, CALIFORNIA

The two dominant California counties for date production are Riverside County, which includes the Coachella Valley (Figure 2), and Imperial County, which includes the Bard Valley. Together these two counties account for over 90 percent of California's total production. A small amount of date production also occurs in San Diego, San Bernardino, and Inyo Counties.

Riverside County began reporting date production in 1925, with 305 tons of fruit produced on 605 acres (Table 3). Average annual production in the county was

Figure 2. The Coachella Valley in Riverside County, California, is the center of date production in the United States. (Donald R. Hodel)

2,651 tons in the 1930s, steadily increasing to 8,341 tons in the 1940s, 13,300 tons in the 1950s, and 15,337 tons in the 1960s. Production declined slightly to 14,800 tons in the 1970s and 11,043 tons in the 1980s, but then increased sharply to 21,244 tons in the 1990s. The overall trend for Riverside County roughly parallels that of the state as a whole, except for the decline in the 1980s. The county's highest annual production was 28,887 tons in 1994. In 2003, production was 17,098 tons on 4,140 acres.

DATE PRODUCTION IN IMPERIAL COUNTY, CALIFORNIA

Imperial County's date production is concentrated in the Bard Valley, across the Colorado River and about eight miles from Yuma, Arizona, although there are a few small areas of production elsewhere in the county. Date production began later in Bard Valley than in Riverside County's Coachella Valley. In 1933 the first commercial date garden was established in the Bard Valley, a 27-acre plot of Khadrawy, Zahidi, and Saidy dates. Medjool largely replaced these varieties beginning in 1944, and is now the predominant variety in the region. Imperial County only began reporting date fruit production in 1947, with 120 tons. Earlier production probably was marketed through outlets in Indio and Phoenix (Winder 1968). Average annual production was 179 tons in the 1950s, steadily increasing to 270 tons in the 1960s, 323 tons in the 1970s, 1,178 tons in the 1980s, and 2,641 tons in the 1990s. Production was 3,737 tons in 2003 on 1,136 acres. Data on bearing acreage were not reported until 1959 (Table 4).

Table 3. Riverside County date production and acreage, 1925–2003

Year	Quantity, excluding by-products (tons)	Bearing and harvested acres	Year	Quantity, excluding by-products (tons)	Bearing and harvested acres
1925	305	605	1965	19,558	3,881
1926	498	573	1966	20,243	3,904
1927	678	700	1967	17,578	4,018
1928	778	600	1968	15,301	4,161
1929	820	700	1969	20,473	4,245
1930	1,491	662	1970	13,772	4,152
1931	1,126	860	1971	16,006	3,823
1932	2,070	1,017	1972	17,000	3,892
1933	2,418	1,217	1973	14,044	3,934
1934	3,113	1,445	1974	13,527	3,888
1935	3,204	1,215	1975	15,623	3,833
1936	3,786	1,500	1976	18,513	3,525
1937	3,507	1,887	1977	11,411	3,499
1938	3,302	2,228	1978	14,634	3,542
1939	2,497	2,548	1979	13,470	3,580
1940	5,671	2,706	1980	13,148	3,579
1941	5,086	2,699	1981	13,522	3,575
1942	7,137	2,715	1982	10,565	3,677
1943	7,993	2,746	1983	9,670	3,677
1944	9,000	3,015	1984	7,600	3,889
1945	3,024	3,049	1985	5,591	3,965
1946	11,450	3,213	1986	8,054	4,195
1947	9,250	3,348	1987	11,467	4,587
1948	11,467	3,433	1988	17,371	5,130
1949	13,335	3,673	1989	13,441	4,486
1950	13,444	3,891	1990	17,878	4,596
1951	15,500	4,062	1991	19,409	4,889
1952	13,250	4,383	1992	20,410	4,692
1953	11,598	4,596	1993	25,268	4,541
1954	10,875	4,404	1994	28,887	4,542
1955	12,183	4,433	1995	18,799	4,476
1956	18,559	4,435	1996	21,359	4,030
1957	12,253	4,493	1997	20,781	4,241
1958	13,617	4,426	1998	22,318	3,848
1959	11,720	3,953	1999	17,335	3,913
1960	15,823	3,944	2000	15,671	3,813
1961	12,165	3,971	2001	14,168	3,507
1962	13,303	4,010	2002	13,004	3,440
1963	14,270	4,008	2003	17,098	4,140
1964	9,655	3,795			

Source: Riverside County Agricultural Crop Reports, 1925–2003.

Aggregated county data do not always agree with statewide figures. In some years the combined totals for Riverside and Imperial counties exceed the figures reported by the State of California. In general, the county data appear to be the more reliable.

DATE ACREAGE BY VARIETY

California has infrequently reported some data on the total bearing and nonbearing acreage of dates by variety (Table 5). Deglet Noor (Figure 3) has decreased from 78 percent of total acreage in 1971 to 70 percent in 1992, while Medjool has slowly but steadily increased from 12 to 19 percent over the same period.

A Riverside County survey completed in 1958 found an estimated 55 acres of Medjool out of a total

Figure 3. Although it has decreased somewhat in recent years, more acreage is devoted to Deglet Noor than to any other variety. (Donald R. Hodel)

Table 4. Imperial County date production and acreage, 1947–2003

Year	Quantity (tons)	Bearing acres*
1947	120	—
1948	100	—
1949	105	—
1950	224	—
1951	166	—
1952	161	—
1953	159	—
1954	153	—
1955	138	—
1956	157	—
1957	127	—
1958	210	—
1959	295	123
1960	254	106
1961	477	106
1962	450	106
1963	132	106
1964	370	123
1965	324	135
1966	229	85
1967	190	90
1968	140	70
1969	130	35
1970	175	50
1971	140	50
1972	456	120
1973	154	140
1974	152	140
1975	348	235
1976	437	282
1977	350	280
1978	480	331
1979	535	382
1980	781	420
1981	781	419
1982	913	450
1983	636	442
1984	1,021	440
1985	1,137	379
1986	1,462	446
1987	1,441	446
1988	1,744	446
1989	1,851	446
1990	1,737	517
1991	2,007	615
1992	2,829	607
1993	3,159	718
1994	2,478	652
1995	2,699	652
1996	2,707	752
1997	2,650	803
1998	3,009	1,003
1999	3,138	1,029
2000	2,826	1,013
2001	3,279	1,028
2002	5,418	1,030
2003	3,737	1,136

*Acreage reporting began in 1959.
Source: Imperial County Agricultural Crop Reports, 1947–2003.

Table 5. California date acreage (bearing and nonbearing) by variety for 1971, 1983, and 1992

Variety	1971 Acreage	1971 % of total*	1983 Acreage	1983 % of total	1992 Acreage	1992 % of total
Barhee	34	<1	53	1	64	1
Deglet Noor	2,907	78	3,867	76	3,910	70
Khadrawy	98	3	102	2	65	1
Medjool	431	12	621	13	1,062	19
Zahidi	201	5	334	6	408	7
Other	63	2	102	2	123	2
Total	3,734	100	5,079	100	5,632	100

*Due to rounding, percentages do not add to 100.
Source: California Fruit and Nut Acreage, California Crop and Livestock Reporting Service, 1993.

of 4,546 bearing and nonbearing acres of dates (Cook 1959), and a follow-up survey in 1968 showed 132 acres of Medjool (Cook 1969). Although no specific data are available, the acreage of Medjool appears to have increased considerably in recent years. In 2002, Imperial County reported 1,030 acres of bearing dates, nearly all of them Medjool and nearly all grown in the Bard Valley.

Commercial growers in Riverside County, particularly in the traditional date areas of the Coachella Valley, also cultivate small numbers of the Barhee, Dayri, Halawy, Khadrawy, Khaisab, Thoory, and Zahidi date varieties, as well as a few other imported varieties; the American commercial varieties are also concentrated in the Coachella Valley. Imperial County grows fewer commercial varieties. Apart from the predominant Medjool plantings, one grower in the Bard Valley cultivates the Dayri, Halawy, and Zahidi varieties on a small scale. Unfortunately, no statistical survey has been done to determine the total numbers of trees or production levels for these lesser commercial varieties.

PRODUCTION TRENDS IN CALIFORNIA

In recent decades the Coachella Valley has seen significant land use changes, as housing tracts, golf courses, shopping malls, and other urban developments have replaced date groves. This shift in land use is common and an increasingly critical issue in most of California's prime agricultural areas. In some instances, date growers have gone out of business altogether while others have moved their operations farther to the south where urbanization pressures are less, using the proceeds from sale of their land (for development) and their palms (for landscaping) to purchase new land and establish new date groves. Many of the affected date groves were quite old and had tall, difficult-to-work palms. Urbanization and relocation of date groves could be viewed as somewhat positive since it allowed and promoted the rejuvenation of bearing acreage within Riverside County. The same urbanization pattern occurred in the greater Phoenix, Arizona, area beginning in the 1960s, and there it ultimately led to the elimination of commercial date orchards because there was nowhere for the date groves to relocate (Johnson, Joyal, and Harris 2002).

At present an estimated 80 to 90 percent of the dates produced in California are sold wholesale, with the remainder sold retail. A number of growers in the Indio area of Coachella Valley once maintained roadside shops to sell some of the dates they produced along with other date-related farm products (including an American innovation, the date milkshake) and to house mail-order operations. When Interstate Highway 10 was opened, bypassing the Coachella Valley business district, through traffic declined, sharply reducing roadside retail sales. Only two of the original roadside shops remain in the Valley. Several growers have continued with retail mail-order sales. The more recent development of Internet sales has added significantly to their trade. A few growers also sell dates at the regular farmers' markets in the Southern California area. Organic date production has also helped the industry to some extent.

From its origins in the mid 1940s, date growing in the Bard Valley has steadily increased. Today more than 1,000 acres are in production, nearly all in Medjool dates. Climatic conditions, fertile alluvial soils, and abundant groundwater make the Bard Valley an ideal date-growing area. Moreover, there is little pressure from urbanization because urban growth in the region has been centered in Yuma, Arizona.

FUTURE DATE PRODUCTION TRENDS

Current estimates show Deglet Noor accounting for 70 to 75 percent of the California date crop, with Medjool accounting for 20 to 25 percent. These estimates support the general observation in the industry that Medjool acreage and production are expanding at the expense of more traditional varieties such as Deglet Noor (Karp 2002). Although the proportion of production represented by Deglet Noor has decreased, it remains the dominant date variety and will likely hold its own in competition with Medjool, for several reasons: an ample supply of Deglet Noor offshoots is available for new and replacement plantings; Deglet Noor can be harvested more easily than Medjool by cutting the entire bunch at one time, thereby reducing labor costs; and as it is a semidry date, the fruit of Deglet Noor can be mechanically pitted.

In general the United States date industry in California and Arizona appears to have a promising future, but a few concerns remain. Although high-quality, whole, domestic dates compete successfully with imported dates in the United States, consumption levels in the United States are very low, only about 2 to 3 ounces per capita annually, based on 2000 data.

The United States both exports and imports dates (Table 6). High-quality, high-value Medjools account for most exports. Overall, exports declined from 7,633 metric tons in 1980 to 3,335 metric tons in 2002, a decrease of 56 percent. Foreign competition in traditional markets such as Europe is partly responsible for this decline. Expansion of the domestic and interna-

Table 6. Annual date exports and imports for the United States in 1980, 1990, 2000, and 2002

Year	Exports (metric tons)	Imports (metric tons)
1980	7,633	6,331
1990	5,245	6,429
2000	3,183	4,667
2002	3,335	4,147

Source: United Nations Food and Agriculture Organization Trade Yearbook, 1981, 1991, 2001, 2002.

tional markets would clearly benefit the date industry in the United States.

Date imports have dropped from 6,331 metric tons in 1980 to 4,147 metric tons in 2002, a decrease of about 34 percent. Imports are typically low-value, bulk-packaged dates used to make confectionery products.

The sale of mature date palms for use in ornamental landscaping, a relatively recent phenomenon, may also play a significant role in the future of the date industry. Most date growers have found it more profitable to sell their date palms for landscape use once they have attained about 25 to 35 feet of trunk height than to continue to grow them for fruit production (W. Geissler, pers. comm.; Karp 2002; D. Young, pers. comm.). If planted at about 50 trees per acre, the palms begin to produce harvestable fruits after three to five years, steadily increasing to full production at 15 to 20 years. Growers can net $500 to $1,000 per acre each year, once the grove reaches full production.

At $1,200 per palm, though, a grower can net about $60,000 per acre once the palms have reached the desired height if the palms are harvested and sold for landscape use after 20 to 25 years (W. Geissler, pers. comm.; Karp 2002; D. Young, pers. comm.). Indeed, much of the recent expansion of date palm acreage in California and Arizona is due as much (if not more) to the demand for palms for landscape use as to demand for fruit production (D. Young and W. Geissler, pers. comm.). Although the removal of still-bearing palms for landscape use obviously decreases a grower's date production, increases in planted acreage may eventually offset this reduction.

Date palm production in California and Arizona follows an unusual agricultural model: it is one of the few crops that actually generates cash income in fruit production to help defray production costs for 20 to 25 years until the palms attain a harvestable, saleable size for landscape use.

The ending of the annual meetings of the Date Growers' Institute in 1979, the closure of the USDA Date and Citrus Station in Indio in 1980, and the lack for more than two decades now of a full-time date palm scientist has left the United States date industry without a strong research base. The California Date Administrative Committee supports research on specific topics through the University of California, Riverside, and the USDA occasionally addresses a date-related research activity, but these do not provide the sustained effort and continuity that the industry will need if it is to continue to be a success in the 21st century.

DATE PALM GERMPLASM RESOURCES IN CALIFORNIA AND ARIZONA

Date palm germplasm is preserved in four formal collections: two in California and two in Arizona (Appendix C). Unfortunately, not all date varieties in the United States are represented in these collections and many (if not most) of the lesser-known and noncommercial varieties have been lost and no longer exist.

The collection at Thermal, California, has the most: 35 varieties, including staminate (male, pollen-producing) and pistillate (female, fruit-producing). The next-largest collection is at Tempe, Arizona, with 32 varieties, then Brawley, California, with 23, and Yuma, Arizona, with 8. Figures for these last two collections are for pistillate palms only.

2

THE DATE PALM

HISTORY AND BOTANY

The date palm is in the genus *Phoenix,* which comprises 14 species (World Checklist of Monocots 2005) of trunked or stemless, solitary or shrubby, pinnate-leaved, armed, pleonanthic, dioecious palms ranging from the Canary Islands and Africa through the Middle East to eastern and southeastern Asia. *Phoenix* palms are distinguished from other pinnate-leaved palms by the upward and lengthwise fold of their pinnae, the reduction of their lower pinnae to spines, and their furrowed seed.

Several species of *Phoenix* are well-known ornamentals in California, including *P. canariensis* (Canary Island date palm), *P. reclinata* (Senegal date palm), and *P. roebelinii* (pygmy date palm). Even *P. dactylifera,* which can be distinguished from these other *Phoenix* species by its large, clustered habit with pinnae arranged in several planes, has become a common ornamental as old palms are removed from date orchards in the Coachella Valley and transplanted into urban landscapes.

The date palm occurs naturally around springs, seeps, streams, and underground sources of water from the Arabia Peninsula to southern Pakistan (World Checklist of Monocots 2005). Early human activity both domesticated the palm and expanded its range west across northern Africa and into Mesopotamia. "Volunteer" dates (escapees from cultivation) occur in almost all date-growing areas, including California.

The date palm has a cylindrical trunk that can grow to a height of 100 feet in 80 to 100 years and bears at its top a crown of spreading, pinnate leaves (Figure 4). Single palms that are maintained and cultivated for dates and have their offshoots removed regularly can live for 100 years or more. Offshoots are new lateral trunks, usually arising at or near the tree's base, but sometimes several feet above the ground. A grower can carefully remove these offshoots and replant them to make new palms. In theory, though, date palm clumps can live indefinitely and become quite old, large, and ever-expanding if you allow them to retain their offshoots. In this case, the clump may never die: it constantly produces offshoots, which in turn produce their own offshoots, continuing in a never-ending progression.

The date palm trunk is 12 to 18 inches in diameter without the spirally arranged, fibrous leaf bases, which can persist for years. After they fall or are removed, the leaf bases reveal a dull brown trunk with diamond-shaped leaf scars four inches high and 10 to 12 inches wide.

Like all monocots, the date palm has adventitious roots, each arising independently from an area of the trunk near the soil called the *root initiation zone.* The expanded mass of exposed roots at the root initiation zone is called the *root boss.*

Figure 4. The date palm has a cylindrical trunk that can grow to 100 feet tall in 80 to 100 years, and bears at its top a crown of spreading, pinnate leaves. (Donald R. Hodel)

Date palms have a limited number of offshoots per palm. The exact number varies by palm variety, and once these have been removed no additional offshoots will arise. When the grower is careful to remove all offshoots, the result is a solitary, single-stemmed palm, the condition found in commercial date orchards and in most landscapes.

The pinnate leaves can number as many as 150 on

one palm, grow as long as 20 feet, and have a life span of three to seven years. They are induplicate, meaning that the trough formed by the fold of a pinna (plural, *pinnae*) points upward when a leaf that is attached to the tree is held out horizontally.

The leaf is divided into three parts: the leaf base or leaf sheath, the petiole, and the blade. The leaf base is the part that attaches to the palm's trunk. Leaf bases are persistent, as much as 12 inches wide, crescent-shaped, and sometimes greenish with maroonish coloring, and have margins with reddish brown fibers encircling the trunk.

The petiole is the part that connects the leaf base to the blade. It is bare and devoid of other appendages and can be anywhere from barely detectable to several feet long.

The blade, which is the largest part of the leaf, holds the pinnae or leaflets. The central stalk of the blade holding the pinnae is the *rachis*. The pinnae nearest the petiole are modified into spines (Figure 5), an identifying character common to all species of *Phoenix*. Spines are solitary or grouped and pointed in various directions. They are rigid and sharp, and they must be removed to allow workers safe access to pollinate, thin fruit and bunches, bag bunches, and harvest fruit (Figure 6). Pinnae can number up to 150 on each side of the blade and are arranged in one to three planes.

Inflorescences (also known as *flower clusters*) and fruitstalks (*infructescences, fruit bunches,* or *fruit clusters*) are held among the leaves and bear the flowers and fruit (Figure 7). The inflorescences and fruitstalks are composed of three parts: the *flower* or *fruit head, the rachis,* and the *peduncle.* The flower of fruit head in turn is composed of the flower- and fruit-bearing strands or *rachillae;* the *rachis,* from which the rachillae arise; and the *peduncle,* the elongated, more or less flattened stalk

Figure 5. The proximal pinnae of date palms, like all species in the genus *Phoenix,* are modified into rigid, sharp spines. (David Karp)

Figure 6. Spines must be removed to prevent injury to workers who must enter the crown of leaves frequently to perform various tasks. (David Karp)

Figure 7. Inflorescences (flower clusters) and fruitstalks (infructescences, fruit bunches, or fruit clusters) of date palms are held among the leaves and bear the flowers and fruit. (Donald R. Hodel)

that holds the rachis and flower or fruit head at one end and attaches to the trunk at the other.

Like all other *Phoenix* species, date palms are dioecious: staminate (male, pollen-producing) and pistillate (female, fruit-producing) flowers grow on separate palms. Pollen from the flowers on staminate palms (Figure 8) must move to the flowers on pistillate palms (Figure 9) for date fruit to form. Staminate inflorescences are shorter than their pistillate counterparts, have a peduncle as long as 20 inches and rachillae as long as 12 inches. Pistillate inflorescences elongate greatly after pollination, and at fruiting time the peduncle is 24 to 60 inches long. When laden with fruit, a pistillate inflorescence is often referred to as *fruitstalk, fruit bunch,* or *infructescence.* A pistillate inflorescence can have as many as 150 flower and fruit rachillae, and each can be as long as 16 inches.

Figure 8. Staminate inflorescences, enclosed in a brownish, thick bract as they emerge, have flowers that produce only pollen. (David Karp)

Figure 9. Once pollinated, pistillate inflorescences will carry the date fruit. (David Karp)

CHARACTERS USED IN DESCRIBING AND DISTINGUISHING DATE VARIETIES

Scientific studies of the date palm required that a means be developed to describe named varieties accurately. The descriptive material that follows is based on Mason (1915a) and Nixon (1950).

For purposes of clarity, the term *adaxial surface* is used here to indicate that side of an organ that is facing the axis that bears it; for example, the upper side of a leaf, petiole, or leaf base. In contrast, the term *abaxial surface* is used to indicate that side of an organ that faces away from the axis that bears it; for example, the lower side of a leaf, petiole, or leaf base. *Distal* refers to the point farthest from the place of attachment; for example, the apex or tip of a leaf. In contrast, *proximal* refers to the point closest to the place of attachment; for example, the base of a leaf. We follow the example of Nixon (1950) and provide metric measurements for the dimensions of the various organs in our descriptions, although we use English measurements elsewhere in this publication. Color designations enclosed in quotation marks ("") are from Ridgway (1912). We provide a key to the major and commercial imported and American varieties of dates in Appendix D.

Trunk

Only large differences in trunk diameter are helpful in distinguishing date varieties. Descriptions and comparisons are based on observations (without specific measurements) of trunks with old leaf bases still attached but moderately pruned, following the standard industry practice for commercial date production in California and Arizona.

Leaves

Color. Leaf color does not vary greatly among date varieties, but descriptions of the various shades of green and varying degrees of glaucous indument are sometimes useful.

Curvature. Varieties vary greatly in leaf curvature, in whether the leaves are stiff or have a slight, moderate, or pronounced curvature, and in whether the curvature is evenly distributed or occurs only at the tip. All of these distinctions often prove valuable in identifying date varieties.

Length. Varieties show characteristic differences in leaf length, but this character is only valuable when observed on palms of more or less the same age growing in similar environmental and cultural conditions. The leaf length increases until palms reach their full production and then decreases again on very old palms. The length measurement is based on the blade, which is measured from the most proximal spine to the tip of the terminal pinna. Leaves with blades less than 11 feet long are described as *short;* leaves from 11 to 14 feet are *medium;* and leaves that are longer than 14 feet are *long.*

Leaf Bases

Most leaf bases are one to two feet long, but overlapping leaf bases and the presence of fibers on the trunk can obscure the exact length. Comparative width, color, and presence of branlike scales or indument (called *scurf*) along the edges are particularly useful. All descriptions of the leaf base refer to the abaxial surface. Fibers attached to the proximal edges of the leaf bases occur in rather distinctive solid sheets or strips in a few varieties, and these are noted if relevant to varietal identification.

Spines

Number. Date varieties vary greatly in their numbers of spines per leaf, although there is some variation within a given variety due to environmental conditions and the age of the individual palm. Fewer than 20 spines is considered *few;* 20 to 30, *medium;* and more than 30, *numerous.*

Spine area. The percentage of the leaf blade length occupied by spines is a helpful character, and is measured from the most proximal spine to the most distal one just before the pinnae. Less than 15 percent is considered *short;* 15 to 25 percent is *medium;* and more than 25 percent is *long.* In some varieties, the transition from spines to pinnae is so gradual that it may not be clear which is the last spine and which is the first pinna. In these cases, the last spine is usually followed by two or three that are intermediate between spines and pinnae. These intermediate types more closely resemble pinnae, though, and are sometimes called *spike pinnae* and distinguished from the other, more normal pinnae or *ribbon pinnae* (Mason 1915a).

Arrangement. The proximal spines usually occur singly while the more distal spines are usually in pairs or occasionally in groups of threes. The pattern of arrangement is often characteristic to the variety.

Thickness and stiffness. Spines of different varieties vary from slender to thick and from weak and pliable to stout and rigid. These terms are relative but sometimes helpful.

Length. Spines less than 4 inches long are considered *short;* 4 to 6 inches long are *medium;* and more than 6 inches long are *long.* The shortest spines are the most proximal and they increase in length distally toward the pinnae. Spine length may be useful for identification, but can vary within a given variety based on the vigor of the individual palm.

Neck. Spines thicken, solidify, and harden where the fold closes at the attachment to the rachis. This thickened, solid, hardened tissue, which usually extends for about 0.4 inch from the end of the fold to the rachis, is called a *neck*. In some varieties, the neck can be 0.8 to 1.6 inches or more long. The neck is generally most pronounced on the more distal spines, the spike pinnae, and occasionally on the few most-proximal pinnae. Because spines of only a few varieties feature a long or conspicuous neck, this feature can be an important identifying character.

Rachis angle. The angle a spine's attachment to the rachis is called the *rachis angle.* Because the rachis angles of single spines differ slightly from those of spines in groups, they are measured separately. This character is only occasionally useful.

A-R divergence. The angle between two spines close enough to form a pair is called the *antrorse-retrorse* or *a-r divergence.* Because a large a-r divergence, when present, is rather striking, this is a useful measurement to know as a supplement to the rachis angle.

Pinnae

Drooping. Varieties vary in the stiffness of pinnae and the extent to which they normally droop or curve downward. Differences may be due to the thickness, length, or width of the pinnae as well as environmental and cultural factors.

Length and width. Length and width of pinnae are important characters. Pinnae less than 24 inches long are considered *short;* 24 to 30 inches are *medium;* and more than 30 inches are *long.* In most varieties, the longest pinnae are the most proximal near the spines, but in a few the longest pinnae are near the middle of the blade. As for width, pinnae less than 1.5 inches wide are considered *narrow;* 1.5 to 1.75 inches are *medium;* and more than 1.75 inches wide are *broad.*

Angles. The angles formed by the pinnae in relation to the rachis or other pinnae vary somewhat within a given variety, but they can still be useful in describing and identifying varieties. Here are the four basic angles to consider:

Valley. The angle formed by ranks of opposite pinnae on the adaxial surface of the blade, measured near midblade and 12 inches from the apex.

Dorsal. The angle formed by ranks of opposite pinnae on the abaxial surface of the blade, measured near midblade and 12 inches from the apex.

Rachis. The range of angles formed by pinnae with the distal extension of the rachis just below midblade.

Apical divergence. The angle between opposite ranks of pinnae with the distal extension of the rachis 12 inches below the apex on the abaxial surface.

Basal spacing index (BSI). The distance between groups of pinnae compared with the linear space they occupy on the rachis, known as the *basal spacing index (BSI),* varies among date varieties (Nixon 1934). Differences are most apparent in the proximal portion of the blade. The BSI, expressed as a percentage, is determined in a three-step process. First, measure the distance between the points of attachment of the most distal pinna of the first and fourth groups of normal (non-spine) pinnae distal of the spine area. Second, measure and sum the space that the second, third, and fourth groups of pinnae occupy on the rachis. Third, divide that sum by your first measurement and multiply by 100 to get the BSI. A BSI of less than 30 percent is considered *low;* 30 to 50 percent is *medium;* and more than 50 percent is considered *high.*

Arrangement. The arrangement of pinnae in groups is distinctive in some varieties and a useful identifying character. For our purposes, a *group* is two or more pinnae relatively close together and separated from other groups by a distance greater than that between pinnae within a group. *Pulvini,* areas of hardened, yellowish white tissue that connect the pinnae to the rachis, are often more or less joined together within a group but are seldom so between groups.

Groups of two or three pinnae are most common; groups of four less common; and groups of five or more rare. Groups are described as *distinct* when they are easily distinguished and *indistinct* when not. Grouping is considered to be *obscured* when the pinnae are closely crowded, as in the distal portion of many leaves.

Classes. The direction in which pinnae are situated in relation to a horizontal plane through the rachis can be useful. *Introrse* means the pinnae are directed in the same plane as the rachis, *antrorse* pinnae are directed upward, above the horizontal, and *retrorse* pinnae are directed downward, below the horizontal. Opposite ranks of antrorse pinnae form a *valley* angle, whereas opposite ranks of retrorse pinnae form a *dorsal* angle, as described earlier. Pinnae are described as *indefinite* when antrorse and retrorse are not very different from introrse, *definite* when the three are moderately different, and *pronounced* when they are strongly different.

Within a group, the proximal pinna usually is antrorse while the distal one usually is retrorse. Pinnae between these two are most commonly introrse but may be antrorse or retrorse. The antrorse-retrorse arrangement of the proximal and distal pinnae respectively in a group is called *regular;* any other situation is termed *irregular* (Mason 1915a).

Fruitstalks

The color and size of the fruitstalks (*fruit bunches* or *infructescences*) and the presence or absence of very small, branlike scales or indument (*scurf*) are often useful for identification. The length and width of the fruitstalks are proportional to the vigor of the palm and vary according to time of emergence, but still they are helpful. Fruitstalks less than 3 feet long are considered *short;* 3 to 5 feet are *medium;* and more than 5 feet are *long.* In some rare circumstances, the length of the rachis (that portion of the fruitstalk at the tip of the peduncle from which the flower- and fruit-bearing rachillae arise) may be important.

Fruit

Fruit develop and mature in stages, and color and other characters change from one stage to another. The following terminology is Iraqi in origin and has become the standard for the date industry.

Kimri is the growing stage during which fruit are green, immature, still developing, and not full sized (Figure 10).

Khalal is the stage at which fruit attain their maximum size and are yellow or red (Figure 11). Color in the *khalal* stage is the most reliable of all fruit characters for classification of varieties.

Rutab is the stage from the time the fruit begin to soften at their tips until they are cured (Figure 12).

Tamar is the fully cured or dried stage of the fruit, when it will not ferment and sour.

Shape. Fruit shape is an important character and is best observed in the *khalal* stage because distinctive differences tend to disappear during ripening. Shape refers only to the symmetrical outline of the fruit unless oth-

erwise noted. The symmetrical view can be determined by making a longitudinal section of the fruit.

Oblong fruit are two to three times as long as they are broad, with little decrease in diameter at either end.

Elliptical fruit are oblong with a curved outline and with ends that are alike in width.

Oval fruit are broadly elliptical with a width that is more than half their length.

Ovate fruit are egg-shaped, with the broader end attached to the rachilla.

Obovate fruit are inverse of ovate: egg-shaped, but with the narrower end attached to the rachilla.

Size. The relative size of fruit can be a useful identifier, although many factors, including pollen (Nixon 1928, 1935, 1936), number of fruit per palm (Nixon 1940), and soil moisture (Moore 1938), can influence it. Measurements are from normal, lightly or moderately thinned fruits. Width refers to a maximum diameter of the fruit seed and flesh in the late *rutab* and early *tamar* stage.

Calyx. Although not part of the fruit, the *calyx* (a collective term for the sepals) adheres to the fruit and is often a characteristic of fruit shape in the *khalal* stage. In some varieties it is appressed against the *corolla* (a collective term for the petals), while in others it may protrude above the corolla. The calyx is described with fruit shape when it is a useful identifier.

The calyx is considered *flattened* or *slightly prominent* when it protrudes less than 1 mm above the corolla; *moderately prominent* when 1 to 2 mm above;

Figure 10. *Kimri*-stage fruit (shown here in Zahidi variety) are green, immature, still developing, and not full-sized. (David Karp)

and *prominent* when protruding more than 2 mm above the corolla. The margin of the calyx may be entire or the triangular tips of the three sepals may be distinct. Enlargement of the fruit may cause the calyx to be slightly or deeply broken, usually between the sepals. When the margin is deeply broken it has a *cleft* appearance.

Perianth is the collective term for the calyx and corolla together as they persist and remain attached to the fruit.

Skin. The skin of date fruit is described as *thin, medium thick,* or *thick.* It may also be characterized as *tender* or *tough.* Differences are relative and have only minor value because many factors, including exposure, season, and handling, can affect the skin. Checking— small linear scars or checks associated with rain or high humidity during the transitional period immediately preceding the *khalal* stage (Aldrich et al. 1946; Nixon 1935)—varies with variety and climatic conditions, and is not always useful in identification.

Flesh. Date varieties have traditionally been grouped into one of three classes—*soft, semidry,* or *dry*—based on the texture or consistency of the fruit under normal ripening conditions. Climate and cultural conditions affect the consistency of the flesh, however, and it is often difficult to determine exact demarcations among the three types. There is so much variation in smoothness of the flesh and *rag* (the fibrous portion of the flesh around the seed) that it is not safe to attempt fine distinctions.

Flavor. Only for a very few varieties is flavor useful for identification. So many factors influence flavor that it cannot be considered a dependable character.

Time of ripening. Varieties are classified as *early, midseason,* or *late.* The following ripening periods are for the Coachella Valley, Imperial Valley, and Yuma, Arizona, area. Ripening in the Salt River Valley, Arizona, usually comes about 2 or 3 weeks later.

Early. Ripening begins around August 15 and lasts for 4 to 6 weeks.

Midseason. Ripening begins around September 1 and lasts for 6 to 10 weeks.

Late. Ripening begins around September 15 and lasts for 8 to 12 weeks.

Seed. The color, size, and shape of seed are of minor importance when it comes to identifying varieties. Pollen has a marked effect on these characters. The

Figure 11. *Khalal*-stage fruit of most varieties are yellow, although a few (as shown here in the American variety Abada) are a striking red. (Donald R. Hodel)

Figure 12. *Rutab*-stage fruit of most varieties are typically amber to brown, as shown here in the American variety Honey. (David Karp)

position of the germ pore (a small, rounded depression marking the location of the embryo) is helpful, especially if it is near the base or the apex.

The central groove or furrow is usually too variable to be of much value. It may be closed or open. If open, it may vary in width and in depth. Furrow descriptions given here are relative and approximate.

Furrow width, determined by the distance between crests (highest points) on each side, is described as follows:

Narrow, one-fourth or less of total width.

Medium wide, between one-fourth and one-half of total width.

Wide, one-half or more of total width.

Depth, determined by its ratio to the dorsiventral thickness of the seed, is as follows:

Shallow, 20 percent or less of thickness.

Medium deep, more than 20 and to 30 percent of thickness.

Deep, more than 30 percent of thickness.

Width and depth are measured at the midpoint between base and apex. The furrow is seldom uniform throughout but typically tends to widen toward the base. Sometimes the apical end of a closed furrow is pitted or has a distinct opening.

IMPORTED DATE VARIETIES

Imported varieties account for nearly all commercial date production in California. Just two varieties, Deglet Noor and Medjool, account for 91 percent of this production. Only two other varieties account for more than 1 percent each: Barhee and Zahidi. We consider these four to be major commercial imported varieties and will discuss them in detail. Another 12 minor commercial imported varieties are discussed here in less detail; these are Amir Hajj, Dayri, Halawy, Hayany, Iteema, Khadrawy, Khisab, Maktoom, Samany, Sayer, Thoory, and Zagloul. Most of the information in this chapter was first published by Nixon (1950). We have made few changes other than to update some information where appropriate. The remaining 144 imported varieties we consider to be noncommercial. Information on those varieties appears in table format in Appendix E.

MAJOR COMMERCIAL IMPORTED VARIETIES

Barhee

Synonyms. Barhi, Berhi, Birhi.

Meaning. Uncertain; Popenoe (1913b) associated the name with hot summer winds (*barh*) at Basra, Iraq, which purportedly have some effect on fruit maturity.

History. Paul Popenoe introduced Barhee, a soft date, in 1913 from Basra, Iraq. In 1902 at Tempe, Arizona, David Fairchild had introduced some offshoots under the name of Berhi; these later proved to be Braim or Buraym, but the name Berhi was still applied to these palms in the Salt River Valley of Arizona as late as 1950. By 1946 there were about 900 Barhee palms, all in California's Coachella Valley.

Distinguishing characteristics. Barhee palms are outstanding for their robust appearance, heavy trunks, and long, stout, slightly to moderately curved leaves with slightly drooping pinnae. The broadly ovate to nearly round *khalal* fruit are usually distinctive because of their more or less abrupt, wedge-shaped taper from the middle to the bluntly pointed apex and in the near-absence of astringency or objectionable tannin flavor.

Palm. Trunk heavy. Leaves "light elm green" with a rather heavy whitish bloom; blade 3.8–4.5 m long; curvature slight to moderate with little increase in flexibility near the apex.

Leaf bases. Broad; green in color, with old leaf bases slightly maroon on the edges; sparse scurf on edges and extending onto blade.

Spines. 28–36; occupying $\frac{1}{5}$ of blade; mostly in groups of two; 2–4 cm long proximally to 8–12 cm long distally; slender to medium heavy; neck 1–2 cm, indefinite; rachis angle 15°–40°; a-r divergence 15°–30°.

Pinnae. Rather stiff with occasional slight to moderate drooping; longest 60–72 cm × 2.4–4.9 cm, widest 55–66 cm × 4.5–5.2 cm (longest and widest occur midblade), terminal 23–40 cm × 2.4–2.6 cm; valley angle 80°–105°, narrowest near midblade, seldom widening more than 10°–20° proximally or distally; dorsal angle 140°–175° proximally to 165°–170° at apex; proximal antrorse and retrorse rachis angles 35°–65°; apical divergence 80°–95°; BSI 30–45%; in groups of two proximally, a few groups of three midblade and distally; classes definite proximally but in midblade and distally antrorse and introrse not well differentiated.

Fruitstalks. Greenish yellow to orange-yellow; slight to moderate scurf proximally; long; heavy; peduncle 2.4 m long, 6.4 cm wide, 2.6 cm thick; rachis 55 cm long; rachillae 142, 34–78 cm long, 2.7–3.7 mm diameter.

Fruit. *Khalal* yellow (Figure 13); *rutab* amber (Figure 14); *tamar* amber to reddish brown; light bloom; broadly ovate to rounded, commonly with a wedge-shaped taper from middle to bluntly pointed apex; 32–37 mm × 23–30 mm; calyx flattened, rounded-triangular, usually three-cleft; skin medium thick, shrinking, with flesh in irregular folds or blistering somewhat; flesh 5–6 mm thick, soft, smooth, translucent, seldom with more than a trace of rag; flavor rich, delicate, exceptionally pleasing in *rutab* stage; late ripening.

Seed. Light brown; oblong, slightly wider above middle, somewhat tapering to blunt apex; 18–23 mm × 8.4–10.5 mm; germ pore central; furrow commonly medium in width and depth.

Notes/culture. Fruit of Barhee are generally regarded as among the best dessert dates because of their excellent flavor. Although the fruit cures and keeps well, its delicate, pleasing flavor (which distinguishes it from other varieties) gradually diminishes under ordinary storage conditions. Fruit tend to be somewhat more susceptible to checking and splitting but they suffer relatively little damage from rot and souring. (*Checking* refers to minute, superficial breaks [checks] in the skin of green dates. It affects the appearance of mature fruit and is caused by high humidity. Breaks may be transverse, longitudinal, or irregular, and vary with the date variety [Nixon and Carpenter 1978]). When checking does occur it is seldom sufficiently pronounced to be considered a serious blemish. Overthinning can result in tough skin, especially if the skin separates from the flesh in ripening and curing. *Khalal* fruit of Barhee are sweet and crisp and are often sold and eaten in this stage. Yields are heavy, averaging 250 to 350 pounds of fruit per palm annually.

Although offshoots of Barhee are large and vigorous for their age, they are few in number, seldom more than six to eight per palm, and are always within 3 feet of the soil surface. Palms showed serious damage in the 1937 freeze (Nixon 1938b) and intermediate damage in the 1949–1950 freezes (Swingle 1950).

Barhee is in all four of the germplasm collections (Appendix C).

Deglet Noor

Synonyms. Deglet Nour, Deglet Nur.

Meaning. *Date of the light* (Popenoe 1913b).

History. The most widely planted and commercially important date variety in the United States, Deglet Noor originated in Algeria in the late 1600s (Swingle 1904) where it soon was recognized as a superior date. A semidry date, its fruit were well known in European markets at the time. It was introduced into the United States at Tempe, Arizona, in 1900, and four years later some of these plants were brought to the Coachella Valley in California. It was soon evident that Deglet

Figure 13. *Khalal*-stage fruit of Barhee are golden yellow, sweet, and crisp, and are often sold and eaten in this stage. (David Karp)

Figure 14. *Rutab*-stage fruit of Barhee are amber and soft. (David Karp)

Figure 15. Deglet Noor is still the most widely planted and commercially important date variety in the United States. (Donald R. Hodel)

Noor produced much better fruit in California than in Arizona. Demand for offshoots rose dramatically and was so strong that commercial growers in California made several large importations of offshoots from northern Africa from 1911 to 1921. Today, Deglet Noor still accounts for about 70 percent total date production in California (Figure 15).

Distinguishing characters. Leaves are long and slightly arched, and have relatively stiff pinnae. The olive-green color and long spine area with numerous spines, a few of them in groups of three, are usually sufficient for identification. The light coral-red *khalal* fruit are distinctive.

Palm. Trunk slender to medium. Leaves yellowish olive; blade 3–5 m long; curvature slight, fairly uniform.

Leaf bases. Narrow to medium broad; green, somewhat glaucous, with age a little maroon color appears irregularly on edges and in the middle near the fiber line; very sparse scurf on edges.

Spines. 40–60; occupying ¼–⅓ of the blade; mostly in groups of two, 8–12 solitary or irregularly proximally, 2–4 groups of three; 4–8 cm long proximally to 16–20 cm long distally; medium stout but more or less variable; antrorse rachis angle 10°–35°, retrorse 10°–35° (-45°); a-r divergence comparable.

Pinnae. Slight to moderate, irregularly drooping and occasionally bending in proximal blade; longest 70–90 cm × 1.4–2.3 cm near spine area, widest 57–72 cm × 3.3–3.9 cm in midblade, terminal 30–50 cm × 1.7–2.3 cm; valley angle 40°–60° proximally to 75°–135° at apex; dorsal angle 130°–170° proximally to 170°–180° at apex; proximal antrorse and retrorse rachis angles 20°–45°, introrse rachis angles 40°–55°; apical diver-

gence 65°–85°; BSI 45–65%; in groups of two and three and a few fours, groups of three predominating in proximal blade; classes usually definite throughout.

Fruitstalks. Greenish yellow to lemon-colored; sparse scurf usually at base; long; slender to medium heavy; peduncle 1.6 m long, 5.3 cm wide, 1.8 cm thick; rachis 39 cm long; rachillae 61, 54–89 cm long, 2.4–3.8 mm diameter.

Fruit. *Khalal* light red; *rutab* amber when soft, brown or straw-colored when dry; *tamar* slightly deeper shades than *rutab* (Figure 16); light bloom; oblong-ovate; 40–50 mm × 20–25 mm; calyx prominent, one- to three-cleft; skin medium thick, adhering to flesh and forming rather coarse wrinkles and folds in curing; flesh 4–5 mm thick, firm, soft, amber except for paler inner zone in which there may be more or less white rag until fruit matures fully; flavor excellent, peculiarly distinctive especially when grown under favorable conditions; late ripening.

Seed. Medium brown; narrowly elliptical; 23–30 mm × 7–9 mm; germ pore central or nearly so; furrow usually closed through the middle, continuing as a slight depression near apex and base with ventral surface more or less flattened.

Notes/culture. Very attractive in appearance and possessing a delicate, distinctive flavor, fruit of Deglet Noor are of firm texture, shrink less in curing, and hold their shape better in packing, handling, and storage than softer varieties (Sievers and Barger 1930). In storage, sugar spots (light-colored or white spots or film on the skin), often a source of considerable trouble with other commercial varieties, seldom mar the fruit's appearance (Barger 1933; Rygg 1942). However, fruit quality is very sensitive to environment, and inferior

Figure 16. Fully dried *tamar*-stage fruit of Deglet Noor are brownish and attractive. (David Karp)

dates are likely to result unless the palms are planted under favorable conditions. Too heavy or too alkaline a soil will often result in inferior fruit.

Fruit of Deglet Noor are especially subject to damage from rain or humid weather. Fruit check (usually in small, transverse, linear scars chiefly near the apex) more easily than those of most other commercial varieties. Darkening and shriveling of the tip (blacknose) usually follows if checking is severe. Severely affected fruit may be nearly worthless. *Khalal* fruit are subject to severe splitting and tearing if they come into direct contact with water. Although losses from fermentation and souring are less than with other varieties, fruit rotting during prolonged periods of humid weather can be serious (Fawcett and Klotz 1932). With few exceptions, Deglet Noor does not appear to be well-adapted to Arizona because of its tendency toward checking (Albert and Hilgeman 1935).

As with all semidry dates, excessive fruit drying occurs during periods of very dry weather or in localities with very low humidity. The best fruit are probably produced where humidity is not too low and ripening occurs after cooler fall weather sets in.

Deglet Noor does best on relatively light soils with sufficient loam or silt to make them retentive of moisture without seriously impeding drainage. A vigorous-growing variety on good soils with proper culture, each palm may be expected to produce 200 to 300 pounds of fruit annually when in full production after 12 to 15 years. Although Deglet Noor seems more susceptible than some other varieties to *Omphalia* root rot (Bliss 1944), which can reduce growth and fruit production, this disease is not a serious problem under optimal soil and cultural conditions (Kenknight and Amling 1947).

Deglet Noor palms seldom produce more than 8 to 12 offshoots, mostly borne at or within 3 feet of the soil surface. Palms showed little damage in the 1937 freeze (Nixon 1938b) and intermediate damage in the 1949–1950 freezes (Swingle 1950).

Because it is still the most common commercial date variety, nearly all of the date palms dug from producing orchards and planted in the urban landscape in California, Nevada, and Arizona are Deglet Noor. However, landscapers consider Medjool and Zahidi palms to have a more attractive crown of leaves and trunk, so those varieties are more highly prized for landscape use.

Deglet Noor is in all four germplasm collections (Appendix C).

Medjool

Synonyms. Medjhool, Medjehuel, Majhul (also Tafilalet, Tafilelt, or Tafilat from the name of the district where it was first grown).

Meaning. Unknown (Popenoe 1913b).

History. For at least two reasons, Medjool merits a more detailed account than some other varieties: it has become a very important variety to the date industry in recent years and its history is somewhat unusual.

Of an estimated 3,000 date varieties grown worldwide (Dowson 1982), Medjool is the most desirable because of its large size, soft flesh, excellent taste, and attractive appearance. All Medjools originate from a single palm in the Bou Denib oasis (Zaid 2002) in the Tafilalet region of Saharan Morocco. Medieval Arabic travelers extolled the very high quality of dates of that region, especially the Medjool, and in the 17th century most of the dates brought to Europe came from Tafilalet. Apparently, at that time, the names *Tafilalet* and *Medjool* were interchangeable (Popenoe 1912, 1973), hence the origin of the name.

The emergence of Bayoud disease, a fatal vascular wilt caused by the soilborne fungus *Fusarium oxysporum,* dealt commercial date growing in Morocco a severe blow around 1870 and has significantly affected the history of Medjool. Bayoud disease destroyed more than 12 million date palms in Morocco and 3 million in Algeria in just one century (Zaid 2002). The disease moves throughout the vascular system, eventually attacking the center bud and killing the palm. Unfortunately,

Medjool is highly susceptible to the disease, and the only effective control measures are to plant Bayoud-resistant varieties (Djerbi 1982, 2003; Zaid 2002) and practice exclusion through quarantines.

There was considerable debate about which standard varieties to introduce as offshoots when commercial date growing was first being discussed and promoted in the United States. Although Medjool was not well known early in the 19th century, Walter Swingle hailed it as one of the best dates grown anywhere and suggested that offshoots be imported from Morocco. Paul Popenoe, who called Medjool the "most famous date in the world," also extolled its virtues (Popenoe 1912). Nevertheless, USDA and commercial importations of offshoots in the early decades of the 20th century, which formed the basis of the California and Arizona date industries, did not include Medjool (Nixon 1971). Medjool seed were imported to California and planted before 1912 and a few offshoots of the variety may have been brought in from Algeria in the 1910s, but details are lacking (Popenoe 1912; Swingle 1904). Regardless, they did not have an important role in subsequent Medjool cultivation.

Walter Swingle was invited to Morocco in early 1927 to take part in a study of Bayoud disease. During a visit to Bou Denib oasis where Medjool originated, Swingle was able to purchase 11 Medjool offshoots, all removed from a single palm growing in a Bayoud-free garden. Shipment of the offshoots to Washington, DC, took five weeks. Upon their arrival the USDA fumigated them and, to ensure they were disease free, required that they be grown in strict isolation under quarantine conditions for several years in a state where no date palms were grown. The southernmost point in Nevada, then without date palms and having a suitable climate, met these conditions. A local Native American farmer near the Colorado River agreed to grow the 11 offshoots, which were planted on July 4, 1927. Three years later the nine palms that survived were producing offshoots and a few fruit bunches. By 1935 the nine surviving original offshoots had produced 64 additional offshoots for a total of 73 palms. Having successfully passed all periodic inspections for Bayoud or other pests or diseases, the palms were released from USDA quarantine in the summer of 1936 and transplanted, without any losses, to the USDA Date Garden in Indio (Swingle 1945; Thackery 1952).

By the early 1940s, it was evident that Medjool represented a promising new commercial date variety for the United States, and the Date Station in Indio began to distribute offshoots to growers in California and Arizona. Medjool is a prolific offshoot producer, a quality that greatly facilitated its establishment in the date industry. Although Medjools were first planted commercially in the Coachella Valley of California, they became especially popular and heavily planted in California's Bard Valley, where they are well adapted to the soils and climate (Figure 17). Today the Bard Valley is the center of Medjool production in California.

Stanley Dillman and Al Collins first brought Medjool dates to the Bard Valley in 1944, using 24 offshoots they had obtained in Indio, California (Berryman 1972; Winder 1968). By 1968 there were about 36 acres of producing and 110 acres of nonbearing Medjools in the Bard Valley (Winder 1968). This increased to about 350 acres in 1971, when there were about 1,700

Figure 17. Medjool is the most common variety in the Bard Valley and is better adapted to the more humid climatic conditions of southeastern California than most other varieties. (Donald R. Hodel)

bearing trees producing about 270,000 pounds of fruit (Berryman 1972). In 2002 there were slightly more than 1,000 acres of Medjool in Bard producing over 5,000 tons of fruit.

In addition to the Bard and Coachella Valleys, Medjools are also grown at China Ranch Date Farm in Tecopa near Death Valley, California. At 36° N latitude, this is the northernmost location for commercial date production in North America. Medjool was also planted experimentally in 1944 on a private farm near Porterville, California, about 175 miles due west of Tecopa in the San Joaquin Valley, but the lower summer temperatures and cool, damp winters at that site were not suited to commercial production of the dates (Nixon 1957).

Commercial planting of Medjool palms began in Arizona near Mesa in 1946 and in Tempe in 1952 (Hilgeman 1972), but because of climatic conditions they performed poorly in both locations. In the late 1990s, the Yuma area, with climatic conditions identical to Bard Valley, began to establish sizable Medjool date gardens. Today more than 1,000 acres of Medjools have been planted south of Yuma, though most have not yet reached bearing age.

Descendants of Swingle's original 1927 importation of Bayoud-free Medjool offshoots from Morocco have not only made possible the expansion of a significant aspect of the California and Arizona date palm industries, they have also been used to establish this variety in countries other than the United States. Offshoots from California are the basis for Medjool production in Israel, where it is the leading date variety and accounts for about 38 percent of that country's estimated date production of 17,000 tons in 2003 (Glasner 2004). California offshoots also provided planting material for commercial date gardens in the Mexican states of Sonora and Baja California Sur.

Medjool is currently the most highly sought after variety and accounts for most of the newly planted date acreage in the United States. The palms are so popular that offshoots sell for up to $300 each, a price that has prompted a rash of thefts of recently planted offshoots from fields in California and Arizona (Hodel and Pittenger 2003a).

Distinguishing characters. The medium to large, broadly oblong-oval to somewhat ovate, reddish brown *tamar* fruit are distinctive. Seed usually have distinctive wings or ridges toward one end.

Palm. Trunk medium heavy. Leaves short to medium long; curvature slight, uniform.

Leaf bases. Medium broad; green, slightly glaucous initially, later becoming somewhat yellowish with a little longitudinal maroon streaking or mottling in center; very sparse scurf on edges.

Spines. 30–38; occupying ¼ of blade; mostly in groups of two with 1–3 groups of three on some leaves; 5–10 cm long proximally to 15–20 cm long distally; medium heavy to stout; neck variable, lacking on many leaves, sometimes 1–2 cm long, indefinite, especially on the middle spine of a group of three.

Pinnae. Slight to moderate drooping with age; longest 70–82 cm × 2.4–3.0 cm, widest 46–54 cm × 4.6–5.2 cm, terminal 23–30 cm × 2.0–3.0 cm; valley angle 50°–65° proximally to 160°–180° at apex; dorsal angle 145°–160° proximally to 160°–180° at apex; apical divergence 55°–75°; BSI 50–75%; in groups of three proximally, occasionally a group of four mid-blade; indistinct above; classes definite throughout, pronounced proximally.

Fruitstalks. Orange-yellow; slight scurf proximally; short to medium long; heavy.

Fruit. *Khalal* orange-yellow with a fine reddish brown stippling; *rutab* amber (Figure 18); *tamar* reddish brown, more or less translucent; moderate to pronounced bloom; broadly oblong-oval to somewhat ovate; 38–48 mm × 26–32 mm; perianth set in a slight depression around the stigmatic scar; calyx moderately prominent, margin rounded-triangular or with one to three slight breaks; skin medium thick, adhering to flesh in curing and forming coarse, irregular wrinkles; flesh 5 to 7 mm thick, moderately soft, very little rag; flavor rich, pleasing; early ripening, just a little ahead of Khadrawy.

Seed. Dark brown; oblong or oblong-elliptical, usually with one or more distinctive wings or ridges; 18–24 mm × 8–9 mm, only 6–8% of total fruit weight; germ pore below middle; furrow variable, closed in middle or narrow and shallow to deep, a little wider at base and apex.

Notes/culture. Fruit of Medjool are highly resistant to damage from humid weather, especially if subjected to late summer water stress. Checking is slight and irregular in character.

Mature fruit of Medjool can be very large and will become so crowded on the rachillae that quality is decreased due to misshapen, damaged, or smaller fruit. Fortunately, Medjool is perhaps the variety most responsive to thinning in terms of fruit size and quality (Codekas et al. 1959). Indeed, relatively heavy fruit thinning is a necessity with Medjool in order to promote uniformity and the extra-large size that commands the premium market price (Furr and Hewitt 1964; Nixon and Carpenter 1978). The higher price received for larger fruit more than offsets the increased costs of thinning.

Thinning is accomplished by removing entire rachillae and also removing individual fruit, rather than removing portions of rachillae. Medjool is the only

Figure 18. *Rutab*-stage fruit of Medjool are amber. (David Karp)

variety for which individual fruit are hand-thinned from rachillae. When fruit are about $^3/_{16}$ inch in diameter, entire rachillae are cut out and removed from the center of the infructescence (*bunch*). The recommended numbers of rachillae that should remain in the fruiting bunch range from 25 to 35 (Winder 1968) to as many as 45 (Nixon 1951). Growers have traditionally thinned fruit to 12 to 16 per rachilla (Furr and Hewitt 1964; Nixon 1956b), although thinning to 16 to 20 fruit per rachilla has resulted in the highest quality fruit (Nixon 1951, 1956b). The pattern for fruit thinning is to remove a fruit when two are opposite each other on the rachilla or when two are side by side (Juan Lukin, pers. comm.). Thinning of individual fruit is best done when fruit are about $^1/_2$ inch in diameter (Winder 1968).

A galvanized spreader ring, commonly inserted into the bunch before the fruit reach the *khalal* stage, keeps the bunch open and allows for ventilation to reduce checking and blacknose. (Blacknose is severe checking near the tip of the fruit that causes it to darken and shrivel [Aldrich et al. 1946; Haas and Bliss 1935; Nixon 1932, 1933].) Some growers place a mesh bag over the bunch to prevent bird damage.

Mature Medjool palms may produce 30 or more inflorescences each season. If all inflorescences are permitted to develop and produce fruit, the palm will bear a large crop that season but an abnormally small crop the next. To maximize and stabilize fruit production, the best practice is to allow only 22 inflorescences, evenly spaced around the crown, to develop and bear fruit each year.

Medjool does have a tendency to drop fruit during picking, a phenomenon called *shattering* (Codekas et al. 1959). As much as 15 to 20 percent of the potential fruit crop can be lost to drop prior to harvest (Nixon 1956b).

Yields have ranged from nearly 150 to 200 pounds of fruit per palm annually (Nixon and Carpenter 1978), although slightly higher yields have been reported in Sonora, Mexico, just south of Yuma, Arizona (Juan Lukin, pers. comm.). Palms showed intermediate damage in the 1949–1950 freezes (Swingle 1950).

Because of its highly desirable fruit, many growers have selected Medjool seedlings for trial and have hybridized other dates with Medjool. Several American date varieties can be traced back to Medjool. Francis Heiny alone made ten such selections. The Medjool parentage is usually clearly evident in the winged or spatulate seeds of these American varieties.

Medjool is in all four germplasm collections (Appendix C).

Zahidi

Synonyms. Zahdi, Zadie, Zaydi, Zehedi, Zaheedy.

Meaning. Uncertain, perhaps *of a small quantity* (Dowson 1939) or *nobility* (Popenoe 1913b).

History. David Fairchild introduced Zahidi, a semidry date, in 1902 from northern Iraq. As with other varieties from that region, though, most of the commercial plantings of Zahidi in the United States are traceable to Paul Popenoe's 1913 introduction. By 1946 there were 183 acres of Zahidi in California (Byrd, Blair, and Phillips 1947); all but 30 acres were in the Coachella Valley, with an estimated 25 acres in Arizona, nearly all in the Salt River Valley.

Zahidi is the most common date variety in northern Iraq and is more widely distributed than any other variety in that country. Some Iraqis believe that Zahidi and Kustawy were the original varieties from which all others were derived.

Some Coachella Valley growers insisted that there

Figure 19. Very fully dried *tamar*-stage fruit of Zahidi are light brown. (David Karp)

were two strains of Zahidi, one producing small fruit and the other medium or, if thinned, fairly large fruit. In a few instances, examination of palms showing differences in size of fruit have shown no consistent differences in leaf characteristics, and it has not been proven that the differences in fruit size are transmitted through offshoots. It may be that with this variety, so widely grown in Iraq, there are seedlings that have been cultivated as Zahidi; this would account for the observed differences.

Distinguishing characters. The Zahidi palm has a characteristic compact crown. The closely set, even-angled pinnae and only slightly curved rachis lend an appearance of solidity to the leaf blades and a stiff formal aspect to the crown. The foliage is glaucous green and the tips of older leaves tend to die back early and take on a light straw color, often distinguishing it from other varieties. The yellow *khalal* fruit have a distinctive, obovate shape.

Palm. Trunk medium heavy. Leaves "jade green" or "light hellebore green"; blade 3.1–4.2 m long; curvature slight, uniform.

Leaf bases. Narrow to medium broad; glaucous green; very sparse scurf sometimes on edges.

Spines. 25–30; occupying ⅟₇–¼ of blade; usually half of them solitary with a rather even, symmetrical arrangement, distally in groups of two; 2–6 cm long proximally to 10–16 cm long distally; slender to medium stout; neck 1–3 cm long, indefinite; rachis angle 40°–75° for single spines, antrorse and retrorse 20°–60°; a-r divergence 25°–50°.

Pinnae. Drooping slightly; longest 55–76 cm × 1.6–3.2 cm and a little distal of spine area, widest 45–56 cm × 3.3–4.3 cm and midblade or a little distally, terminal 18–26 cm × 0.7–1.8 cm; valley angle 60°–90° proximally to 110°–140° at apex; dorsal angle 120°–150° proximally to 170°–190° at apex; proximal antrorse and retrorse rachis angles 25°–50°, introrse 40°–70°; apical divergence 60°–80°; BSI 30–70%; in groups of two and a few threes, occasionally a group of four, regular proximally, becoming indistinct distal of midblade; classes indefinite distal of midblade.

Fruitstalks. Orange-yellow; sparse to moderate scurf proximally; medium long; medium heavy; peduncle 1.2 m long, 6.0 cm wide, 2.1 cm thick; rachis 24 cm long; rachillae 81, 45–67 cm long, 2.2–4.4 mm diameter.

Fruit. *Khalal* yellow, some with more or less fine brown stippling; *rutab*, softer portions light brown, drier basal portions faded yellow or straw-colored; *tamar*, softer portions reddish brown, drier portions little changed from *rutab* (Figure 19); light bloom; obovate; 34–40 mm × 23–25 mm; calyx prominent, rather abruptly elevated, one- to three-cleft; skin rather thick and tough, tending to adhere to flesh, loosening little in curing; flesh 4–5 mm thick, firm, of smooth consistency in softer fruit with very little rag, drier flesh more or less fibrous and becoming rather hard; flavor not outstanding, lacking in delicacy; ripening midseason.

Seed. Grayish brown; oblong and slightly spatulate; 21–25 mm × 7.8–8.7 mm; germ pore central or nearly so; furrow variable, commonly closed in middle, opening slightly at apex and base but sometimes moderately deep and wide and rather uniform.

Notes/culture. Fruit of Zahidi are of the semidry type and lend themselves to handling with less expense than most other varieties. Growers do not need to pick individual fruit but can wait until they are about three-quarters ripe and then cut the entire bunch, finishing the less mature fruit in maturation chambers and softening the drier ones with steam hydration. Hydration may actually improve the flavor of Zahidi, which is not true of Deglet Noor, but the flavor of Zahidi, while generally considered rather distinctive, is not outstanding. The seed is large in proportion to the fruit.

Although fruit of Zahidi do not sour readily, they are more subject to splitting from rain and rot from humid weather than such varieties as Halawy and Khadrawy. Checking occurs in irregular lines toward the apex.

The Zahidi palm is vigorous and has a reputation for being relatively hardy and resistant to disease and drought. Yields are heavy, 200 to 300 pounds of fruit per palm per year. The palm commonly produces 15 to 25 offshoots per year, a few of which may be 12 to 20 feet above the ground. Zahidi palms showed less damage than nearly all other varieties in the 1937 freeze (Nixon 1938b) and the 1949–1950 freezes (Swingle 1950).

Zahidi is in the Thermal, Brawley, and Tempe germplasm collections (Appendix C).

MINOR COMMERCIAL IMPORTED VARIETIES

Amir Hajj

Synonyms. Amir Haj, Mirhage.

Meaning. *Leader of the pilgrimage.*

History. A soft date and long reputed to be one of the best dates of northern Iraq (Fairchild 1903; Popenoe 1913b), Amir Hajj originated at Mandali oasis, which is said to be the source of more good date varieties than any other locality in Iraq. Commonly referred to as "the visitor's date," it is generally presented to guests and much of the crop was exported as gifts from the people of Mandali to their friends. Nixon introduced Amir Hajj in 1929 and most of the offshoots were planted in the lower Rio Grande Valley of Texas. Later plantings were made at Indio, California.

Distinguishing characters. Distinctive features of this palm include its vigorous appearance (Figure 20); the rather long pinnae, with those in the proximal portion of the blade drooping and producing the somewhat tangled look of the crown; and the tendency for fiber to occur in more or less solid bands across the leaf bases with looser, upward projections near the bud.

Palm. Trunk heavy. Leaves "light cress green"; medium to long; curvature slight to moderate, fairly uniform.

Leaf bases. Medium broad, green, sparse scurf on edges extending onto petiole and proximal rachis.

Spines. 14–26; occupying $\frac{1}{8}$–$\frac{1}{5}$ of blade; variable, but mostly in groups of two; 2–6 cm long proximally to 12–18 cm long distally; slender to medium stout; lacking neck.

Pinnae. Drooping moderate to pronounced, especially in proximal blade and on older leaves; 50–74 cm × 1.5–3.9 cm, terminal 28–37 cm × 1.7–2.1 cm; in groups of two proximally, a few groups of three midblade, occasionally a group of four distally, regular.

Fruitstalks. Greenish yellow, sometimes orange-tinted; medium long; medium heavy.

Fruit. *Khalal* yellow; *rutab* amber; *tamar* reddish brown with moderate bloom lending bluish cast; oblong-oval; 32–36 mm × 20–22 mm; skin thin, tender, shrinking with flesh; flesh soft, smooth, caramel-like, little or no rag; flavor delicate, rich, intermediate in sweetness between Maktoom and Barhee; ripening midseason.

Seed. Grayish brown; oblong; 18–22 mm × 8–9 mm; furrow deep, narrow.

Notes/culture. Reportedly best adapted to sandy soils, Amir Hajj is a vigorous grower that produces moderately large yields. *Rutab* fruit are nearly too soft to be handled easily but if left on the palm they cure into attractive, high-quality dates.

Figure 20. Vigorous in appearance and reputed to be the best date from northern Iraq, Amir Hajj has much promise because of its large yields and high-quality fruit. (Donald R. Hodel)

A promising variety, the relatively small fruit size may be its most conspicuous drawback. While there is some checking, early observations have shown that there has been relatively little damage from fruit rot. Palms produce about 12 offshoots each.

Amir Hajj is in the Thermal and Brawley germplasm collections (Appendix C).

Dayri

Synonyms. Dairee, Dairi.

Meaning. *The monastery date* (Popenoe 1913a), or *of Dayr,* a place in southern Iraq so named because of its location at a convent (*dayr*) (Dowson 1939).

History. A semidry date introduced from Basra, Iraq, in 1913 by Paul Popenoe, Dayri was accidentally mixed with other varieties but as the palms came into bearing they attracted attention because of the size, color, and quality of their fruit. Erroneously grown under the names Awaidi and Hawazi, two rare varieties from southern Iraq, the palm was finally and positively identified by Nixon as Dayri in 1928 (Nixon 1934). By 1946 there were about 200 Dayri palms in the Salt River Valley of Arizona and 225 in California, more or less evenly split between the Coachella and Bard Valleys.

Distinguishing characters. Distinctive features include the rather dark, glaucous green leaves with somewhat pronounced curvature near the tip; an open

Figure 21. Dayri has an open and airy aspect to its crown of leaves. (Donald R. Hodel)

appearance to the center, and an airy aspect to the entire crown (Figure 21) resulting from the long spine area and moderately long leaves with narrow pinnae, the latter drooping and in well-spaced groups just proximal of the spine area; the fibers tending to be drawn in tight, solid bands 3–4 cm wide across the green leaf base; and the dark, nearly black fruit.

Palm. Trunk slender to medium heavy. Leaves near "dark cress green"; blade 3.6–3.8 m long; curvature moderate, increasing rapidly toward the tip.

Leaf bases. Narrow to medium broad, rather flattened; green, somewhat glaucous, sometimes maroon on edges of older leaves; scurf lacking or very sparse.

Spines. 20–26; occupying ¼ or more of blade; mostly in groups of two; 4–8 cm long proximally to 12–18 cm long distally; slender, stiff; neck 1–3 cm long.

Pinnae. Drooping moderately proximally; longest 48–76 cm × 1.8–3.9 cm, terminal 23–30 cm × 1.5–2.3 cm; mostly in groups of two with a few groups of three, regular.

Fruitstalks. Greenish yellow to pale orange-yellow; medium long; slender to medium heavy.

Fruit. *Khalal,* dull red over orange-yellow background; *rutab,* soft fruit deep brownish red with darker shades at base, dry fruit pale purplish brown; *tamar* showing little change except some darkening of soft fruit that become nearly black at the base with light bloom lending a purplish cast; oblong to oblong-elliptical with rounded apex and more or less oblique base; 36–45 mm × 19–23 mm; skin tough but tending to shrink with flesh; flesh in soft dates tender, amber, and translucent, little rag, in dry dates tough and hard, especially near the base where it is whitish; flavor of soft dates good, rather distinctive; flavor of dry dates often marred by slightly disagreeable aftertaste; ripening midseason.

Seed. Light brown; oblong-elliptical; 28–31 mm × 8.5–10.0 mm; furrow closed in the center but gradually opening toward base and to a lesser extent near apex; ventral surface of seed characterized by a faint depression on either side of the furrow and appearing as a line parallel to it.

Notes/culture. Like all semidry dates, fruit of Dayri may be soft, dry, or intermediate in consistency depending on climate, culture, and handling. Dry and soft fruit of Dayri are so different that they may easily be taken for two different varieties. The dry fruit are inferior, being unattractive in appearance and lacking in flavor when compared to those of Thoory and Deglet Noor. In contrast, softer fruit, when grown under favorable conditions, are large and attractively shaped and have pleasing flavor and firm skin that shrinks from the flesh, making it better adapted to commercial handling than some soft dates of better quality. The very dark, soft fruit can be used to give contrast in combination packs with lighter-colored dates.

With the exception of Sayer, no other variety is so tolerant of drought and adverse soil and climatic conditions as is Dayri. Unfortunately, when grown on light, sandy, alkaline soils with insufficient irrigation, Dayri produces a larger proportion of drier, inferior fruit. Excellent quality fruit and high yields, 240 to 300 pounds of fruit per palm annually, have been achieved when growing Dayri on heavier soils with ample irrigation. Because it is often grown on poor soils with insufficient irrigation in Iraq, Dayri is not highly esteemed there.

Fruit of Dayri have sustained moderate damage from splitting during rain, but rot and souring have been no more problematic than in other Iraqi varieties. In fact, Dayri has produced its best fruit in the Coachella Valley during seasons of humid weather—it tends to develop a larger proportion of dry and inferior fruit in dry seasons.

Dayri produce an abundance of small but easily rooted offshoots, commonly 15 to 20 per palm, typically within 3 feet of the soil surface. Palms showed little damage in the 1937 freeze (Nixon 1938b) but sustained intermediate damage in the 1949–1950 freezes (Swingle 1950).

Dayri is in the Thermal, Brawley, and Tempe germplasm collections (Appendix C).

Halawy

Synonyms. Halawi, Hallawi, Hellawi.

Meaning. *Sweet* (Dowson 1939).

History. A soft date, David Fairchild introduced Halawy in 1902 from northern Iraq. As with other varieties from that region, though, most of the commercial plantings of Halawy in the United States are traceable to Paul Popenoe's 1913 introduction. By 1946 there were about 45 acres of Halawy in California and 30 acres in Arizona.

One of the most common varieties in southern Iraq, Halawy was generally regarded as one of the best varieties shipped in quantity from Basra, Iraq, to European and American markets.

Distinguishing characters. A combination of characters distinguishes Halawy, including the open center of the crown; deep, somewhat glaucous, green, medium-long leaves with moderate and rather uniform curvature; broad, stiff pinnae; and light-amber-colored, wrinkled ripe fruit.

Palm. Trunk medium heavy. Leaves "pois green" to "hellebore green" with moderate bloom lending a glaucous, slightly bluish cast; blade 3.3–4.0 m long; curvature slight to moderate, rather uniform.

Leaf bases. Medium broad, green, little maroon on edges of old leaf bases; slight to moderate scurf on edges extending onto proximal petiole and rachis.

Spines. 18–33; occupying $\frac{1}{4}$–$\frac{1}{3}$ of blade; mostly solitary, typically with 4–8 spines in rather distant groups of two distally; 2–4 cm long proximally to 12–16 cm long distally; variable, medium stout on young and vigorous palms, slender and weak on old ones.

Pinnae. Stiff; 40–60 cm $\times$ 2.5–4.8 cm, terminal 23–43 cm $\times$ 1.5–3.0 cm; in groups of two and three, the former predominating proximally.

Fruitstalks. Orange-yellow, sometimes orange-tinted; medium long; medium heavy.

Fruit. *Khalal* yellow (Figure 22); *rutab* light amber (Figure 23); *tamar* golden brown with moderate bloom; oblong with rounded apex; 35–45 mm $\times$ 17–20 mm; skin thin, shrinking with flesh in irregular wrinkles

Figure 22. *Rutab*-stage fruit of Halawy are amber while the same variety's *khalal*-stage fruit are yellow. (David Karp)

with only slight blistering; flesh soft, caramel-like, translucent, with little or no rag, amber outer zone, light golden inner zone; flavor very rich, sweet, distinctive; early ripening.

Seed. Grayish brown; narrowly oblong, apex bluntly rounded but usually with a short point, often slightly wider above the middle, frequently with a very slight constriction between this point and the base; 18–28 × 6.5–9.0 mm; furrow deep, varying from shallow to deep.

Notes/culture. Halawy is unusually problem-free and has several attributes making it a good-quality date. Fruit cure and keep exceptionally well. Rain and humid weather have caused relatively little loss of fruit, although high humidity can cause severe checking.

The primary drawback to Halawy is from a merchandising standpoint: its fruit shrivel and wrinkle in ripening more than those of any other date. In very dry seasons or when irrigation is deficient, especially on light soils, fruit of Halawy tend to shrivel before reaching maturity and some retain dry, hard areas around the base. The best fruit of Halawy have been produced on heavier soils with ample irrigation.

Palms of Halawy are moderately vigorous, yielding 150 to 200 pounds of fruit per palm annually and typically 10 to 15 offshoots per palm, some of these 6 feet or more above the soil surface. Offshoots have small connections and are readily removed and rooted. Palms are relatively tender, and showed severe damage in the 1937 freeze (Nixon 1938b) and 1949–1950 freezes (Swingle 1950).

Halawy is in the Thermal, Brawley, and Tempe germplasm collections (Appendix C).

Hayany

Synonyms. Hayani.

Meaning. From *Hayan,* a man's name (Martius 1823)—probably the owner of the original palm.

History. A soft date introduced by Fairchild in 1901 from Egypt, Hayany was the most extensively planted and important commercial date variety in Lower Egypt. Nearly all U.S. commercial plantings of Hayany were confined to Arizona's Salt River Valley, where about 50 acres were grown in 1946. Some palms of this variety were erroneously carried under the name Birket el Haggi in Arizona in the early 1900s (Mason 1927).

Distinguishing characters. Hayany palms have an airy, graceful crown of moderately arched leaves with long, drooping pinnae and long, slender spines. The larger spines have a distinctive neck as long as 3–5 cm; *khalal* fruit are deep red.

Palm. Trunk slender. Leaves "hellebore green" with moderate bloom lending glaucous cast; blade 3.1–4.0 m long; curvature moderate, fairly uniform.

Leaf bases. Medium broad or less; green; old leaf bases have maroon on edges; sparse scurf occasionally on edges.

Spines. 19–33; occupying $\frac{1}{10}$–$\frac{1}{5}$ of blade; about $\frac{1}{3}$ of them in rather distant groups of two but commonly appearing to be solitary; 2–6 cm long proximally to 18–23 cm long distally; slender, stiff; neck 3–5 cm long.

Pinnae. Drooping more or less pronounced; 80–100 cm × 1.4–4.2 cm, terminal 34–45 cm × 2.5–3.5 cm; in groups of two with a few groups of three.

Fruitstalks. Orange-yellow, sometimes with a greenish cast; medium long; medium heavy or more.

Fruit. *Khalal* deep red; *rutab* and *tamar* nearly black,

Figure 23. *Rutab*-stage fruit of Halawy are soft, elongate, and amber. (David Karp)

light bloom lending purplish cast; oblong-elliptical; 45–55 cm × 22–28 mm; skin medium thick, rather tough, often separated from flesh; flesh soft, watery, rather coarse with considerable rag; flavor mild, but lacking in distinctive quality; early ripening.

Seed. Grayish brown; oblong, often with a slight dorsal longitudinal depression extending up from base to about 1/4>> the length of seed; 23–32 mm × 8.0–9.5 mm; furrow narrow, shallow to rather deep.

Notes/culture. Fruit of Hayany are distinctive because they have high moisture content and do not readily cure. Although fruit can be successfully held in low-temperature storage (Hilgeman and Smith 1938), it is generally marketed as a fresh date. The fruit appeal to consumers who prefer dates that are less sweet.

Fruit of Hayany can be picked in the *khalal* stage. Although still hard and deep red with considerable tannin, after a few days they usually soften and become more palatable. *Khalal* fruit are sweet and crisp at harvest, however, so Hayany dates are often sold and consumed in this stage.

A heavy and dependable producer (King, Beckett, and Parker 1938), Hayany commonly yields 250 to 300 pounds of fruit per palm every year. Unfortunately, the fruit are very susceptible to damage from checking and souring from rain or humid weather. Hayany palms produce an unusually high number of offshoots, 30 to 40 per palm, and these can be 3 to 6 feet up the trunk. The palms are relatively hardy, showing the least damage in the 1937 freeze (Nixon 1938b) and 1949–1950 freezes (Swingle 1950).

Hayany is in the Thermal, Tempe, and Yuma germplasm collections (Appendix C).

Iteema

Synonyms. Itima, Itime, Ytima, Yatimeh.

Meaning. *The orphan* (Popenoe 1913b).

History. Although Swingle introduced Iteema, a soft date, in 1900 from Algeria, most of the commercial plantings of this date can be traced to introductions Bernard Johnson made from Algeria in 1913 and 1914. By 1946 there were about 23 acres of Iteema in the Salt River Valley, Arizona, but only about 150 of the palms in California.

Although Iteema was said to be very common and popular in Algeria (Kearney 1906), literature accounts indicate that there were problems in identifying (and applying this name to) date palms throughout northern Africa and that several other varieties were probably grown under this name in that region.

Distinguishing characters. Distinctive features of this variety include its long, moderately arched leaves with pinnae and numerous spines somewhat appressed to the rachis. Also, young palms have loose, papery folds and sheets of fiber conspicuous at the leaf base around the bud. Although this character becomes less pronounced with age, it is often apparent even on old palms from the adherent fiber on the trunk below the crown.

Palm. Trunk medium heavy. Leaves "jade green"; blade 4.0–4.8 m long; curvature moderate, rather uniform.

Leaf bases. Medium broad, strongly rounded; green, old leaf bases have traces of maroon mottling on their edges; very sparse scurf sometimes on edges and extending onto petiole and proximal rachis.

Spines. 45–50; occupying $\frac{1}{5}$–$\frac{1}{4}$ of blade; mostly in groups of two with one or two groups of three; very poorly developed or nearly lacking proximally to 8–14 cm long distally; stout; neck 1 cm long.

Pinnae. Rather stiff at apex with moderate drooping proximally; 65–91 cm × 1.7–4.5 cm, terminal 14–27 cm × 1.0–2.6 cm; in groups of two and three, a few groups of four, and an occasional group of five.

Fruitstalks. Orange-yellow; medium long; medium heavy.

Fruit. *Khalal* yellow with more or less fine brown stippling; *rutab* light brown; *tamar* reddish brown with moderate bloom; oblong-obovate, apex broadly pointed; 38–50 mm × 20–26 mm; skin medium thick and a little tough, tending to leave flesh in curing; flesh soft, melting, with only a trace of rag, more or less translucent; flavor very sweet, rather cloying; ripening midseason.

Seed. Brown; oblong to oblong-elliptical, usually slightly wider above middle; 21–24 mm × 7–9 mm; furrow closed in midportion or narrow and shallow, slightly wider and deeper toward base and open or narrowly pitted for about 2 mm at apex.

Notes/culture. Although an attractive date of excellent flavor, Iteema did not prove popular in California's Coachella Valley: its fruit are susceptible to checking, fermenting and souring, and dropping, and tend to become puffy with rain and humid weather. Fruit will not cure on the palm except in the most favorable seasons, and heavy losses have occurred in damp weather. Cooler, drier weather results in better fruit and less souring. Fruit can be harvested just as they begin to soften slightly at the tip and ripening can be completed in maturation rooms with artificial heat.

Iteema palms produce 8 to 12 medium-size offshoots, usually within 3 feet of the soil surface. They are difficult to root and transplant, and Albert and Hilgeman (1935) recommended that growers leave offshoots on the mother palm 1 to 2 years longer than is typical for other varieties and then plant and grow them for at least 2 years in the nursery.

Iteema grows and produces best on light, well-drained soils with ample water and fertilizer. The palms showed moderate damaged in the 1937 freeze (Nixon 1938b) but only a little damage in the 1949–1950 freezes (Swingle 1950).

Iteema is in the Tempe germplasm collection (Appendix C).

Khadrawy

Synonyms. Khadrawi, Khadhrawi, Khudrawee.

Meaning. *Green* (Dowson 1939), referring either to the often-characteristic greenish cast of the just-softening fruit or to the rather distinctive bright green petiole/rachis and leaf base (Popenoe 1913b).

History. Although Fairchild introduced Khadrawy, a soft date, in 1902 from Basra, Iraq, the commercial plantings can be traced to introductions made by Popenoe in 1913. By 1946 there were 183 acres in California (Byrd, Blair, and Phillips 1947) and about 105 acres in Arizona.

In Iraq there are two varieties that go under the name Khadrawy. The more common and important one, known simply as Khadrawy, is from the Basra region in the southern part of the country and is one of the primary varieties of that area. The other variety is from Baghdad and northern Iraq and goes under the name Khadrawy of Baghdad (see Appendix E). This is an entirely different palm, although its fruit is very similar to that of the Basra Khadrawy.

Distinguishing characters. One of the most distinctive of all varieties, the Khadrawy palm is slow in vertical growth and old palms are conspicuously smaller than those of any other commercial variety. The moderately arched leaves, relatively few in number, have more or less flattened blades with short, rather stiff, closely and evenly set pinnae. The terminal pinnae are slightly longer than those immediately proximal, lending a somewhat fishtail appearance to the leaf tip.

Palm. Trunk medium heavy. Leaves "jade green" with less glaucous cast than most other varieties; blade 3.0–3.8 m long; curvature moderate, increasing slightly distally.

Leaf bases. Narrow; bright green; very sparse scurf on edges sometimes extending onto proximal petiole and rachis.

Spines. 15–25; occupying $^{1}/_{7}$–$^{1}/_{4}$ of blade; mostly solitary but about six to eight commonly in distant groups of two; 1–2 cm long proximally to 10–12 cm long distally; slender, medium stiff.

Pinnae. Stiff except for slight drooping in proximal blade; 42–63 cm × 1.7–4.4 cm, terminal 24–35 cm × 1.5–1.9 cm; in groups of two with a few groups of three midblade.

Fruitstalks. Greenish yellow to orange-yellow; medium long or slightly less; slender to medium heavy.

Fruit. *Khalal* yellow (Figure 24); *rutab* amber (Figure 24); *tamar* reddish brown with light bloom; oblong-elliptical to oblong-oval; 33–40 mm × 20–24 mm; skin medium thick, sometimes blistered but shrinking more or less with flesh; flesh soft, melting, becoming caramel-like, little or no rag; flavor pleasant, rich, without being cloying; early ripening.

Seed. Dark brown, becoming lighter near base and there near "pinkish buff"; narrowly oblong or oblong-elliptical; 20–25 mm × 7.7–9.8 mm; furrow variable, sometimes narrow and shallow to deep or sometimes closed in the middle and opening toward base and apex.

Notes/culture. The Khadrawy has several desirable qualities. The fruit cure and keep well with a little less shrinkage and deterioration in appearance than most soft dates. Rain and humid weather have not seriously damaged the fruit. Checking is seldom pronounced. *Khalal* fruit will split rather deeply upon contact with water, however, and fruit left on the palm to cure, as is common practice in the Coachella Valley, will sometimes rot severely in the centers of heavy bunches after rain. Even then, though, fruit of Khadrawy are still not as severely affected by rots as are those of Deglet Noor.

Khadrawy palms perform satisfactorily on a wide range of soils and their relatively small size slightly decreases the cost of harvesting and other operations where workers must ascend into the crowns. At full bearing, yields range from 100 to 150 pounds of fruit per palm annually. Palms typically produce 15 to 20 offshoots each, some as high as 6 feet up the trunk. Palms showed severe damage in the 1937 freeze (Nixon 1938b) but only moderate damage in the 1949–1950 freezes (Swingle 1950).

Khadrawy is in all four germplasm collections (Appendix C).

Khisab

Synonyms. Khasab.

Meaning. *Fruit-bearing palm* (Dowson 1939); *the abundant producer* (Popenoe 1913b).

History. A mediocre, soft date, Nixon introduced Khisab in 1929 from Basra, Iraq, where it was valued because of its very late ripening date. It was first grown at Welasco and Winter Haven, Texas, and at Indio, California.

Palm. Trunk heavy. Leaves medium long; curvature moderate.

Leaf bases. Medium broad; green with a yellowish cast on the proximal edges, sparse scurf sometimes on edges.

Spines. Numerous; occupying $^{1}/_{5}$ or less of blade; nearly all in groups of two, very few in groups of three or four; stout; neck 1 cm long.

Pinnae. Stiff on younger leaves but slight to moderate irregular drooping with age; medium length and

Figure 24. Bunches of Khadrawy fruit, some in *khalal* stage (yellow) and some in *rutab* stage (amber), make an attractive display. (David Karp)

width; in groups of two and three with a few groups of four.

Fruit. *Khalal* deep red (Figure 25); *rutab* dull, reddish brown; *tamar* nearly black with light bloom lending a slight purplish cast; rounded-oval; medium size; skin tough, separating from flesh in blisters; very late ripening.

Seed. Oblong-elliptical; furrow shallow and broad or nearly closed by flattened ventral surface.

Notes/culture. The very late ripening of Khisab fruit is a distinct disadvantage in Texas, although in Arizona and California it is valuable because the palm carries fruit during the late fall and winter tourist seasons after other varieties have been harvested. Fairchild (1903) stated that Khisab ripened in August in Iraq and was one of the heaviest bearers in the region, yielding as much as 450 pounds of fruit per palm annually.

"Nagal" (see Appendix E), a somewhat similar variety primarily grown in Arizona, may actually prove to be the same variety as Khisab.

Khisab is in the Tempe germplasm collection (Appendix C).

Maktoom

Synonyms. Maktum.

Meaning. Uncertain, possibly *kept* (Dowson 1939), *hidden* (Bonavia 1885), or *the bitten* (Popenoe 1913b).

History. Although Fairchild introduced Maktoom, a soft date, to the United States in 1902 from northern Iraq, the commercial plantings can be traced to introductions Popenoe made in 1913. By 1946 a little over 6 acres were planted to Maktoom in California and about 23 acres in Arizona.

Figure 25. *Khalal*-stage fruit of Khisab are a striking, deep red. (David Karp)

Considered one of the best soft dates of the Baghdad region, Maktoom occurs throughout northern Iraq but is nowhere numerous. Dowson (1923) reported it to be rare in southern Iraq.

Distinguishing characters. A combination of characters distinguishes Maktoom palms. Leaf color is more whitish than that of most other varieties. The long, narrow proximal pinnae are closely set in widely spaced groups. Leaf bases on leaves retained until 3 or 4 years old usually have a dark maroon coloring contrasting with the white bloom on the adaxial surface of the petiole.

Palms of Kush Zebda (see Appendix E) closely resemble Maktoom, but Kush Zebda's lack of dark coloring on old leaf bases and lack of moderate to heavy scurf on the proximal petiole and edges of leaf bases distinguish it from Maktoom.

Palm. Trunk medium heavy or more. Leaves nearly "deep grape green" but somewhat glaucous; blade 3.5–4.1 m long; curvature slight to moderate, fairly uniform.

Leaf bases. Medium broad; light glaucous green, some maroon coloring usually near fiber line on second or third year; moderate to heavy scurf on edges extending onto proximal petiole and rachis.

Spines. 10–18; occupying $\frac{1}{8}$–$\frac{1}{6}$ of blade; variable, usually two to four groups of two among distal spines; 1–2 cm long proximally to 8–16 cm long distally; slender to medium heavy; neck less than 1 cm long.

Pinnae. Drooping moderate to pronounced in proximal blade and on older leaves; 58–86 cm × 1.7–4.1 cm, terminal 23–39 cm × 1.1–2.3 cm; mostly in groups of two with many groups of three, rarely a group of four.

Fruitstalks. Orange-yellow; medium long; medium heavy.

Fruit. *Khalal* yellow; *rutab* amber; *tamar* reddish brown with moderate to rather heavy bloom lending a purplish cast; oblong with broadly rounded base and apex or oblong-oval; 30–40 mm × 22–28 mm; skin medium thick, tending to shrink with flesh in irregular folds with some loosening and blistering; flesh soft, melting, slightly mealy, becoming caramel-like, seldom more than a trace of rag; flavor mild, delicately sweet; late ripening.

Seed. Dark brown; oblong, usually tapering from middle to somewhat pointed apex; 20–25 mm × 8.0–8.4 mm; furrow closed, commonly slightly "cracked" toward base and usually with a slight opening at apex.

Notes/culture. Fruit of Maktoom are usually fairly large and have a mild, delicate, appealing flavor. They are remarkably resistant to checking and splitting from rain or humid weather. Although *khalal* fruit are very little affected by rot, they will often sour if left on the palm to cure. In storage, fruit are more susceptible to spoilage than varieties like Deglet Noor and Khadrawy; for this reason, Maktoom appears best adapted to handling as a fresh date.

A moderately vigorous grower, the Maktoom palm performs satisfactorily on a wide range of soils. At full bearing, yields range from 175 to 225 pounds of fruit per palm annually. Palms typically produce 15 to 20 offshoots each, usually within 3 feet of the soil surface. Palms showed severe damage in the 1937 freeze (Nixon 1938b) but only moderate damage in the 1949–1950 freezes (Swingle 1950).

Maktoom is in the Tempe germplasm collection (Appendix C).

Samany

Synonyms. Samani, Samiani, Rashedi.

Meaning. Said to be named for a village (Popenoe 1913b).

History. Introduced from Egypt in 1925 by Mason, Samany was regarded as one of the best dates of Lower Egypt although it was confined to just two localities, Edku and Rashid (Rosetta) (Mason 1915b). Samany is a large, soft date, attractive in appearance in the *khalal* or early *rutab* stages but disappointing in the later stages of ripening at which most of the fruit would be consumed in the United States. Oasis Date Gardens currently has one palm of Samany that produces over 200 pounds of fruit annually, most of which is sold in the *khalal* stage (K. Corral pers. comm.).

Distinguishing characters. The large, oblong-ovate, yellow to dull amber to brown fruit distinguish Samany from other imported varieties.

Palm. Vigorous. Trunk medium heavy. Leaves long; strongly recurved in distal third.

Leaf bases. Fiber light brown, rather loose, more or less in solid bands 3–4 cm wide across leaf bases.

Spines. 42–46; occupying $\frac{1}{4}$ of blade; mostly in groups of two, a few groups of three; 3–6 cm long proximally to 14–18 cm long distally; medium stout; neck 2–3 cm long.

Pinnae. Drooping slight; 56–78 cm × 1.5–4.8 cm, terminal 25–35 cm × 1.4–1.8 cm; mostly in groups of two, a few groups of three.

Fruitstalks. Greenish yellow; medium long; slender to medium heavy.

Fruit. *Khalal* yellow, commonly rather sparsely stippled with red; *rutab* dull amber; *tamar* brownish; oblong-ovate; 50–60 mm × 25–35 mm; skin thin, tender; flesh soft but rather coarse with considerable rag; flavor mildly sweet in early *rutab* stage, becoming insipid later; ripening midseason.

Seed. Light brown; oblong; 28–36 mm × 9–11 mm; furrow irregular, mostly deep and medium broad, sometimes closed or nearly so at middle.

Notes/culture. Fruit of Samany have a tendency to sour easily, even in favorable weather when other varieties have cured and kept well. Checking occurs in short longitudinal lines near the apex.

Samany is in the Thermal germplasm collection (Appendix C).

Sayer

Synonyms. Sayir, Saiar, Sai, Ista'amran, Usta Imran, Sta'amran, Sa'amran, Sambran, 'Amran.

Meaning. *Common, found everywhere* (Bonavia 1885).

History. A soft date, Fairchild introduced Sayer from Iraq in 1902. Popenoe imported additional offshoots in 1913 but few survived. By 1946 there were about 300 palms in Arizona's Salt River Valley but only a few palms in California.

According to Dowson (1923), Sayer was the most common variety in southern Iraq. The fruit were extensively exported and were the principal variety among the cheap, soft, imported dates first sold in the United States.

Dowson (1923) recognized two strains of Sayer. In one, *khalal* fruit are solid yellow, while in the other they are yellow flushed with pink. Only the latter type is grown in Arizona and California.

Distinguishing characters. When viewed from a distance, the crown of the Sayer palm appears somewhat flattened above. The compact crown has succeeding whorls of leaves that are vertically closer than in any other common variety. The moderately curved leaves have a short spine area and the old leaf bases usually have dark maroon coloring near the fiber line.

Palm. Trunk medium heavy. Leaves "jade green"; blade 3.1–4.0 m long; curvature moderate, only slightly increasing in flexibility distally.

Leaf bases. Medium broad; glaucous green, old ones usually with dark maroon coloring near fiber line; sparse scurf on edges, little tufts of fiber commonly adhering to edges of petiole just proximal of spines.

Spines. 12–24; occupying $^{1}/_{10}$–$^{1}/_{5}$ of blade; in groups of two distally, solitary proximally; 2–4 cm long proximally to 8–16 cm long distally; slender to medium heavy; lacking neck.

Pinnae. Fairly stiff but proximal, older pinnae sometimes drooping slightly to moderately; 52–71 cm × 1.8–4.1 cm, terminal 19–28 cm × 1.5–2.2 cm; in groups of two with only a few groups of three.

Fruitstalks. Orange-yellow; medium long; slender to medium heavy.

Fruit. *Khalal* yellow, usually with a little red near base; *rutab* amber; *tamar* reddish brown with moderate bloom; oblong to oblong-oval; 40–48 mm × 19–22 mm; skin thick, tending to adhere to flesh in curing; flesh soft, curing smooth, caramel-like, very little rag; flavor good but lacking in delicacy; ripening midseason.

Seed. Grayish brown; narrowly oblong; 22–29 mm × 7.2–8.4 mm; furrow variable, most common shallow and medium broad through middle, slightly more open at apex, a little deeper and wider toward base, but often fairly uniform throughout and moderately deep.

Notes/culture. Sayer is not highly regarded in Iraq, probably because most of the palms are grown on marginal land that is too salty, too dry, or otherwise unsuitable for other crops. In Arizona and California, though, Sayer has been grown under more favorable conditions and has performed well.

Although somewhat lacking in delicacy of flavor, the fruit of Sayer have several positive attributes. They are attractive in appearance and hold their shape fairly well for a soft date. Fruit on a single bunch ripen more uniformly than those of many other varieties. Losses from rain or humid weather have been small and the fruit hold up well in storage and do not sour readily.

The primary drawback of Sayer is the tendency for many peduncles of fruitstalks on older palms to break during the early stages of development, usually near the base and before the broken portion has emerged from the leaf axil and its enveloping fiber. This condition is known as *crosscut* because the breakage point has the appearance of having been transversely cut partway through. When the breakage is less than half the width of the peduncle, the fruit will continue to develop but the dates on those rachillae in line with the cut will be dwarfed; others within the same bunch will develop normally. The cause of this disorder is unknown.

Yields of Sayer range from 175 to 200 pounds of fruit per palm annually. Palms typically produce 10 to 15 offshoots each, usually within about 3 feet of the soil surface. Palms showed moderate damage in the 1937 freeze (Nixon 1938b) but only slight damage in the 1949–1950 freezes (Swingle 1950).

Sayer is in all four germplasm collections (Appendix C).

Thoory

Synonyms. Thuri, Tsuri, Thauri.

Meaning. *The bull's date* (Popenoe 1913b).

History. Although Swingle introduced Thoory, a dry date, from Algeria in 1900, the commercial plantings can be traced to a few offshoots that accompanied large Deglet Noor importations that Bernard Johnson brought in 1912. By 1946 there were about 580 Thoory palms in California's Coachella and Imperial Valleys and only a few in Arizona's Salt River Valley, where they had been confused with another, inferior dry date, Mesh Degla (see Appendix E).

Distinguishing characters. The Thoory palm is a robust grower with slightly arching, yellowish green

Figure 26. *Khalal*-stage fruit of Thoory are yellow while those in the *rutab* stage are light brown. (David Karp)

leaves having broad, stiff, rather crowded and irregularly set pinnae. Thoory resembles Mesh Degla but the latter differs in that it has conspicuous scurf on its fruit-stalks. The size and color of the ripe fruit distinguish Thoory from other dry varieties.

Palm. Trunk medium heavy. Leaves "jade green" to "cedar green," with very light bloom; blade 3.75–4.75 m long; curvature slight, fairly uniform, with only a little increase in flexibility distally.

Leaf bases. Broad, heavy; yellowish green, old ones with slight maroon on edges and somewhat mottled in center; sparse to moderate scurf on edges extending onto petiole and proximal rachis, weathering away on older leaves.

Spines. 20–27; occupying $^1/_{10}$–$^1/_5$ of blade; mostly in groups of two; 1–3 cm long proximally to 10–16 cm long distally; slender to medium stout; neck variable, usually only 1–2 cm long but sometimes 3–4 cm long.

Pinnae. Harsh, stiff; 56–76 cm × 2.3–6.3 cm, terminal 32–56 cm × 2.6–3.4 cm; mostly in groups of two with about half as many groups of three and a few groups of four; closely and irregularly set, rather wide pinnae lend crowded appearance.

Fruitstalks. Orange, without scurf; short to medium long; heavy.

Fruit. *Khalal* yellow; *rutab* and *tamar* (Figure 26) light brown in softer portions commonly near apex, sometimes a little reddish, drier portions light grayish brown, light bloom; oblong with rounded apex; 37–45 mm × 20–23 mm; skin thick and tough; flesh firm to rather hard and brittle; flavor of a pleasing, nutty character; late ripening.

Seed. Grayish brown; oblong-elliptical; 21–25 mm × 7.5–9.0 mm; furrow variable, commonly closed in middle with ventral surface somewhat flattened, a shallow depression of increasing width toward base and a slight, narrowly pitted opening of medium depth and width at apex.

Notes/culture. Undoubtedly Thoory is the best of the dry dates introduced into the United States. Its large size, attractive appearance, and delicate flavor make it appealing to those who desire dry dates. It has not been too widely planted, though, because of a market preference for softer fruit.

When fully ripe, fruit lack the astringency so characteristic of many dry dates popular in the Old World. Fruit of Thoory have sometimes been harvested a little too early because dates in the fresh ripe and cured stages differ little in appearance. Fruit have been only slightly damaged from rot and souring after rain. Splitting from rain immediately prior to ripening has been confined to rather small ruptures near the stem end.

A vigorous and robust grower, the Thoory palm has done well on light soils in the Coachella Valley of California. Yields range from 200 to 250 pounds of fruit per palm annually. Thoory typically produce few offshoots, seldom more than six to eight per palm, and rarely more than 3 feet above the soil surface. Palms were among those least damaged in the 1937 freeze (Nixon 1938b) and 1949–1950 freezes (Swingle 1950).

Thoory is in all four germplasm collections (Appendix C).

Zagloul

Synonym. Zaglul.

Meaning. Said to be the name of an Egyptian family (Mason 1915b).

History. Mason introduced Zagloul, a soft date, from Egypt in 1922.

Distinguishing characters. The Zagloul palm somewhat resembles the Hayany, but the definite, pronounced neck of the longer and more numerous spines of Hayany make it easy to distinguish. Also, fruit of Hayany have less rag than those of Zagloul. The petiole of the Zagloul palm is unusually long, 20 to 24 inches, and the fiber is rather loosely arranged around the leaf bases with some solid strips or folds around the bud and younger leaves.

Palm. Very tall but more slender than Samany. Leaves 3.4–3.7 m long; curvature slight, stiff (Mason 1915b).

Spines. Spine area short, 38–46 cm long; mostly solitary; 3.8 cm long proximally to 15.2 cm long distally; slender (Mason 1915b).

Pinnae. Drooping slight; 46–66 × 2.5–3.8 cm, terminal 30–38 cm long; mostly in groups of two, a few groups of four (Mason 1915b).

Fruitstalks. Short; heavy (Mason 1915b).

Fruit. *Khalal* deep red; *rutab* and *tamar* deep dull brown, nearly black with light bloom; oblong, sloping somewhat to a bluntly pointed apex; large size; flesh soft; ripening midseason.

Seed. Oblong; furrow variable, usually medium wide and shallow.

Notes/culture. Nixon (1950) reported that the performance of Zagloul was disappointing. A large, soft fruit similar to that of Hayany, Nixon noted that the fruit sour more easily than those of Hayany and are subject to checking, which can cover the entire fruit. Nixon went so far as to conclude that there was nothing to justify its further planting. Brown Date Garden currently grows Zagloul but has only a few bearing palms, each producing about 10 pounds of fruit annually. Brown is considering abandoning and destroying the palms (Ted Fish, Jr., pers. comm.). *Khalal* fruit of Zagloul are sweet and crisp and have been sold and consumed in this stage.

AMERICAN DATE VARIETIES

Early in the 20th century, before offshoots of imported varieties were available in quantity, American growers propagated date palms by planting large quantities of seed. Many of the resulting seedling date palms were so closely spaced and so poorly tended that they never reached normal flower or fruit production stages. During the campaign to eradicate the especially harmful *Parlatoria* date scale (*Parlatoria blanchardii),* the United States Department of Agriculture encouraged and actually underwrote the removal and destruction of these seedling palms because their jungle-like growth made pest detection and treatment difficult. Thousands of seedlings were destroyed during this period, some of which might have had horticultural merit had there been an opportunity to thoroughly evaluate them. Fortunately, a few growers cared for and developed their seedling plantings rationally and judiciously, selecting promising variants and culling out the others. Over the course of several decades, the propagation of the best of these seedling palms resulted in the identification of a number of new date varieties.

Roy Wesley Nixon studied these new American date varieties for several years. He developed botanical descriptions of them as a basis for evaluating any unusual or distinctive characters and recorded information about their origin and culture. Unfortunately, promising horticultural varieties are sometimes lost before they can be thoroughly evaluated. During the more than four decades Nixon studied date palms, several varieties originating in the Coachella Valley and in small-scale production disappeared and were lost from horticulture, including Cleopatra, Desert Queen, and Desert Nuggett. In 1955, 10 acres of one variety, Sphinx, were in commercial production near Phoenix, Arizona. This Sphinx acreage was subdivided for housing tracts but some of the palms were left in place on residential lots and the fruit harvested (Johnson, Joyal, and Harris 2002). Low economic returns for date growers since World War II accelerated the loss of some of the promising but little-known American varieties.

Several of the American date varieties were in limited production, but none competed on a large scale with the main commercial varieties. With several varieties in limited production and others, some with potentially promising merit, disappearing from the annals of horticulture, Nixon saw the need to chronicle and catalog the American date varieties. Information about the history and origin, identification, plant and fruit characteristics, and culture of these varieties was and still is invaluable.

In his unpublished manuscript, Nixon described 40 American varieties, 34 from California and six from Arizona, and divided them into two groups, major and minor, depending on the number of palms of each variety in production (Nixon 1955). He treated varieties for which 1 acre or more (about 50 palms or more) were in full production in 1955 as major and covered them in detail; others he treated as minor and gave them less space, descriptions being limited mostly to the fruit, with some notes on origin and any outstanding characters of the plant.

In this book we present Nixon's American varieties alphabetically in two groups, commercial (9 varieties) and non-commercial (31 varieties), depending on whether the fruit are offered for sale (Table 7). We feel this grouping is more appropriate in that it reflects the importance of these varieties to the commercial date industry in the United States. Our arrangement of commercial and non-commercial groups more or less corresponds to Nixon's major and minor groups, but with a few exceptions. Our commercial group includes three varieties—Abada, McGill's No. 1, and T-R—that Nixon had treated as "minor varieties," and for this reason we have relatively scant information about them. Conversely, five of our non-commercial varieties—Amber Queen, El Toby, Haziz, Lindy, and Sultany—were among Nixon's "major varieties," and so are covered here in more detail. Many of the non-commercial varieties have probably been lost and no longer exist. Fortunately, germplasm collections in California and Arizona (Appendix C) include eight American varieties: Abada, Empress, Haziz, Honey, Peggy Ann, Sphinx, Tabarzal, and T-R.

There seems to have been a significant decrease in

Table 7. American date varieties

Commercial	Non-commercial		
Abada	Amarillo	Frau Damke	Peggy Ann
Blonde Beauty	Amber Queen	Haziz	Plum Date
Brunette Beauty	Andrate	Heiny	Precioso
8-Ball*	Benji*	Heinyana	Rubidor
Empress	Black Lucifer	Humbarger	Shields 6-7
Honey	Black Medjool	Laguna	Smoky
Mariana*	Black Royal	Lindy	Sultany
McGill's No. 1	Deglar	Long Yellow	Sweetheart
Sphinx	Desert Dew	Maktoom No. 1	Wieman's Date
Tabarzal	El Toby	Malakai	Wonderful Black
T-R	Ford Date	Metzler's Black	

*New commercial American variety not included in Nixon's unpublished manuscript.

interest in American varieties since Nixon completed his unpublished manuscript in 1955. A few, mostly smaller growers in California are experimenting with new varieties, some of which have not yet been formally described and named. Pato's Dream Date Gardens in Thermal has 12 bearing palms of Mariana, which has fruit similar to its possible parent, Halawy. China Ranch Date Farm near Tecopa just south of Death Valley offers a China Hybrid and Black Beauty, which are harvested from offshoots taken from seedling dates planted in the 1920s. Ben Laflin, Jr., in Indio has one tree each of two dark-fruited varieties, Benji and 8-Ball, selected from Barhee/Medjool crosses he had made.

Although we consider Mariana and 8-Ball to be commercial varieties because they are offered for sale, we have chosen to include them in an addendum since they appeared well after Nixon developed his manuscript of American varieties.

Whether any of these commercial American varieties will achieve a significant market share remains to be seen. Several of the varieties are promising, but most are not likely to compete with the dominant, standard imported varieties. Growers may, however, be able to exploit specialized market niches or marketing channels. Selling *khalal* fruit of some varieties, making use of different-colored dates in attractive multi-variety packs, or offering dates in farmer's markets and by mail order or the Internet are several strategies that might help to ensure the market permanence of some of these American varieties.

Abada

History. D. G. Sniff discovered Abada growing wild in a riverbed in Brawley, California, in 1936. The varietal name was created out of the first names of Sniff and his wife, Abby and Dana. Most of the offshoots were sold to Sunnipalms, a commercial date garden near Indio, California, where there were 18 palms of Abada in production in 1955. At this writing, there are 25 producing Abada palms in three or four date gardens. The fruit are subject to checking in long, narrow, transverse apical lines. The palm resembles that of Deglet Noor (Figure 27) but has shorter fruitstalks. The fruit has been generally well received by those who have tasted it.

Carpenter (1979) noted that Abada had characters desirable for date breeding, including attractive, glossy black fruit with a frost-like bloom. Abada is in the Thermal and Tempe germplasm collections (Appendix C); in the Tempe collection it is referred to as Black Abada.

Fruit. *Khalal* deep red; *rutab* and *tamar* (Figure 28) black with moderate to heavy bloom lending a pur-

Figure 27. Resembling Deglet Noor in habit, the American variety Abada, shown here with fruit bunches protected by paper coverings, has desirable characters for breeding, including attractive and flavorful black fruit. (Donald R. Hodel)

plish cast; 48–52 mm × 20–22 mm; calyx moderately prominent, slightly to deeply broken, red; skin medium thick, coarsely wrinkled, blistering slightly; flesh soft and melting; rag slight and unobjectionable; flavor very sweet and rather cloying; early ripening.

Seed. Narrowly oblong-elliptical; 29–33 mm × 7.0–7.5 mm; germ pore central or slightly above; furrow closed or very narrow and shallow.

Blonde Beauty

History. Blonde Beauty is believed to have originated from Deglet Noor seed that J. H. Northrop planted on his property near Indio. B. F. Shields later acquired Northrop's property, selected and propagated the variety, and had 4 acres and at least 50 palms in production in 1955. Today there are 433 producing palms at Shields Date Gardens in Indio.

Distinguishing characters. The palm has some resemblance to Deglet Noor but differs in its wider pinnae, the terminal pinnae being shorter and having a larger apical divergence, and its shorter fruitstalks with a medium to heavy scurf on the proximal portion.

Palm. Trunk medium heavy. Leaves long; curvature slight, evenly distributed or increasing distally.

Leaf bases. Medium broad; yellowish green with slight maroon mottling; scurf slight.

Spines. 49, occupying about ¼ of blade; mostly in groups of two and occasionally three on some leaves; 6–20 cm long; medium heavy, stiff; neck 1 cm long, indefinite; rachis angles of paired spines 15–40°; a-r divergence 25–50°.

Pinnae. Drooping slight; longest 78–84 cm × 1.8–2.8 cm; widest 55–58 cm × 4.1–4.7 cm; terminal 23–28 cm × 2.4–2.7 cm; valley angle from 45° proximally to 90° at apex; dorsal angle from 160° proximally to 180° at apex; rachis angles in proximal blade 20–60°; apical divergence 105–145°; BSI 40–60%; in groups of two and three, distinct except apically; classes pronounced proximally, definite throughout.

Fruitstalks. Greenish yellow; scurf medium to heavy proximally; medium long; medium heavy.

Fruit. *Khalal* yellow; *rutab* amber; *tamar* reddish brown; oblong-elliptical; 45–50 mm × 20–24 mm; calyx moderately prominent, margin moderately broken to nearly three-cleft; skin tender, wrinkled with some blistering sometimes resulting in longitudinal folds; flesh soft and melting; rag unnoticeable; flavor pleasing and "taffy-like"; ripening midseason.

Seed. Medium brown; narrowly oblong-elliptical; 23–33 mm × 7–9 mm; germ pore variable but near middle; furrow variable-closed and ventral surface somewhat flattened, or medium to broad and shallow.

Notes/culture. Fruit are purportedly not very susceptible to checking. When present, checking occurs in short, transverse lines, mostly in the apical half of the fruit. Blonde Beauty has produced an average of 16 or more offshoots per palm, some of which are carried 6 to 8 feet up the trunk. Shields Date Gardens (Indio, California) markets Blonde Beauty as its exclusive variety.

Brunette Beauty

History. Like the preceding, Brunette Beauty is believed to have originated from Deglet Noor seed that J. H. Northrop planted on his property. B. F. Shields later acquired Northrop's property, selected and propagated the variety, and had 4 acres with at least 50 palms in production in 1955. Today there are 153 producing Brunette Beauty palms at Shields Date Gardens in Indio.

Distinguishing characters. Brunette Beauty is distinguished by the moderate to pronounced drooping of its pinnae; an abrupt spine-to-pinnae transition; spine midribs with moderate scurf; and fruitstalks with very heavy scurf.

Figure 28. *Khalal*-stage fruit of the American variety Abada are a striking, deep red, while those in the *rutab* and *tamar* stages are black. (Donald R. Hodel)

Palm. Trunk slender. Leaves short; curvature slight, increasing distally.

Leaf bases. Narrow; green with yellowish cast on some, slight maroon mottling longitudinally in center on oldest.

Spines. 41, occupying about ¼ of blade; in groups of two with one or two groups of three; 8–18 cm long; medium heavy, medium stiff; neck lacking; rachis angles of paired spines 20–65°; a-r divergence 20–30°.

Pinnae. Drooping moderate to pronounced; longest 86 cm × 1.9 cm; widest 68 × 3.8 cm; terminal 28 cm × 1.2 cm; valley angle from 55° proximally to 110° at apex; dorsal angle from 170° proximally to 180° at apex; rachis angles in proximal blade 25–40°; apical divergence 65°; BSI 25–40%; in groups of two and three and occasionally four, distinct except near apex; classes definite except near apex.

Fruitstalks. Greenish yellow; very heavy scurf proximally; long; slender.

Fruit. *Khalal* medium red (perianth red except for yellow dominating calyx); *rutab* and *tamar* nearly black; oblong-elliptical; 42–47 mm × 20–23 mm; calyx

Figure 29. A selection that grew from Thoory seedlings, the American variety Empress, shown here with fruit bunches protected by paper coverings, has desirable characters for breeding, including attractive, large, high-quality fruit with a distinctive, rich flavor. (Donald R. Hodel)

prominent, margin slightly to moderately broken; skin medium tough; flesh soft; rag slight and unobjectionable; flavor good; late ripening.

Seed. Medium brown; narrowly oblong-elliptical; 30–33 mm × 6.6–7.3 mm; germ pore above middle; furrow closed and ventral surface often flattened in middle, or sometimes medium-wide and shallow.

Notes/culture. Shields Date Gardens (Indio, California) markets Brunette Beauty as its exclusive variety.

Empress

History. Coming from the same seed lot as T-R, Empress is a Thoory seedling that E. K. Davall planted on his property near Cathedral City, California, in 1916 or 1917. Although Davall sold his property and original plantings, which later became known as the Wonder Date Garden, he had previously started a new garden about a half-mile to the north and had transferred his best seedling palms there, including Empress. Davall had about 15 acres with at least 50 palms in production in 1955. The Davall date orchard presently has about 10 acres and 500 palms.

Distinguishing characters. The palm resembles Thoory in some respects (Figure 29) but the leaflets are narrower and a little less stiff and have a wider rachis angle. The orange-yellow fruitstalks and fruiting rachillae are exceptionally heavy. Fruit are large, oblong, somewhat pointed at the apical end, and may retain dry, light buff- or straw-colored areas at the base. Cured fruit are sometimes oblong-triangular.

Palm. Trunk medium to heavy. Leaves medium long; curvature slight, increasing distally.

Leaf bases. Medium to wide; green, a little maroon at base of oldest leaves; scurf lacking.

Spines. 34–38, occupying slightly more than ⅕ of blade; about one-fourth of spines clustered near base, others in groups of two and sometimes three; 7–17 cm long; broad, stiff; without neck; rachis angles of paired spines 15–25°; a-r divergence 15–30°.

Pinnae. Drooping slight; longest 70–82 × 1.8–2.6 cm; widest 44–47 cm × 4.1–5.0 cm; terminal 30–34 cm × 3.0–4.3 cm (the latter "doubled"); valley angle from 65° proximally to 100° at apex; dorsal angle from 170° proximally to 180° at apex; apical divergence 65–75°; BSI 40–80%; in groups of two, three, and occasionally four, distinct except a few distally of middle; classes definite throughout.

Fruitstalks. Orange-yellow; without scurf; short; heavy.

Fruit. *Khalal* yellow; *rutab* amber (Figure 30); *tamar* reddish brown ("chestnut" or "mahogany red"); oblong, usually sloping from maximum diameter a little below middle to bluntly pointed apex; 46-54 mm ×

Figure 30. *Khalal*-stage fruit of the American variety Empress are yellow while those in the *rutab* stage are amber. (David Karp)

21–24 mm; calyx moderately prominent, three-parted; skin medium, tending to shrink with flesh in course wrinkles; flesh soft; rag slight, unobjectionable; flavor pleasing; ripening midseason.

Seed. Medium brown apically, buff at base; oblong-elliptical; 22–30 mm × 8.0–9.4 mm; germ pore variable, usually slightly basal to near center; furrow shallow and narrow, or closed and opening a little toward apex and base.

Notes/culture. According to David Davall, Empress produces only one to five offshoots per palm and they are more difficult to grow than most other varieties. Full-bearing palms are said to produce 300 to 400 pounds of fruit annually, but occasional poor fruit set has lowered this amount. The fruit are rated more rain tolerant than those of Deglet Noor and checking occurs in irregular, apical lines. The Empress palm flowers later in the spring than does the Honey palm. Empress is in the Thermal germplasm collection (Appendix C).

Carpenter (1979) noted that Empress had characters desirable for date breeding, including attractive, high-quality, large fruit with a distinctive, rich flavor. Ream (1975) reported it was used to hybridize with the imported varieties Dayri and Tadala.

Honey

History. Like Empress, Honey is a Deglet Noor seedling that E. K. Davall planted on his property near Cathedral City, California, in 1916 or 1917. Although Davall sold his property and original plantings, which later became known as the Wonder Date Garden, he had previously started a new garden about a half-mile to the north and had transferred his best seedling palms there, including Honey. Davall had about 15 acres with at least 50 palms in production in 1955. Presently only about 90 palms of this variety exist in three or four date gardens.

Distinguishing characters. Honey is similar to Deglet Noor in its leaves and long greenish yellow fruit-stalks, but differs in its wider pinnae. Fruit are oblong-oval and smaller, slightly broader, and softer than those of Empress.

Palm. Trunk medium heavy (Figure 31). Leaves long; curvature slight, increasing distally.

Leaf bases. Medium wide; green, a trace of maroon mottling on oldest leaf bases; scurf variable, slight to rather heavy.

Spines. 36, occupying slightly more than $\frac{1}{5}$ of blade; in groups of two; 6–19 cm long; medium stout, stiff; neck 1 cm long, indefinite; rachis angles of paired spines 20–45°; a-r divergence 20–25°.

Pinnae. Drooping slight, a few proximally may bend at base; longest 73–80 cm × 2.7–3.0 cm; widest 51–54 cm × 4.7–5.2 cm; terminal 31–35 cm × 2.5–2.8 cm; valley angle from 55° proximally to 120° at apex; dorsal angle from 150° proximally to 175° at apex; apical divergence 85°; BSI 55–60%; in groups of two proximally, mostly in groups of three and four distally, mostly indistinct distally of middle, partly due to coalescence of groups; classes definite.

Fruitstalks. Greenish yellow; scurf lacking or inconspicuous; long; medium heavy.

Fruit. *Khalal* yellow; *rutab* amber ("raw sienna" to "amber brown") (Figure 32); *tamar* reddish brown ("chestnut" or a little deeper); oblong-oval; 40–47 mm × 21–23 mm; calyx moderately prominent, margin rounded with one or two slight breaks; skin medium thick; flesh soft and melting; rag slight, unobjectionable; flavor mild and pleasing; ripening purportedly over a broad time period, beginning before Empress and continuing later.

Seed. Medium brown, slightly grayish; oblong-elliptical, maximum diameter usually a little distal of middle; 23–28 mm × 8–9 mm; germ pore middle or slightly above; furrow shallow, narrow to medium width, or closed and pitted at apical end, opening a little near base.

Notes/culture. Honey is said to have yields comparable to Deglet Noor and to produce a similar number and character of offshoots (6–12 per palm). Its fruit are slightly less susceptible to damage from rain, though, and checking occurs in short, transverse lines in the apical portion.

Honey is in the Thermal and Tempe germplasm collections (Appendix C).

McGill's No. 1

History. A seedling of unknown parentage, D. E. McGill selected and propagated this variety on a ranch about 2 miles east of Mecca, California. The original palms were destroyed when the property later changed

hands, but some offshoots had already been sold, and in 1955 about 30 young palms were growing but not yet bearing in a planting 6 miles south of Indio, California. Today only three producing McGill's No. 1 palms exist. Fruit bear some resemblance to those of Khalasa and their susceptibility to rain damage is similar to that of Deglet Noor. The palm itself is suggestive of Kustawy.

Fruit. *Khalal* yellow; *rutab* amber; *tamar* translucent, reddish brown; oblong oval, very slightly obovate; calyx moderately prominent, margin slightly broken; flesh soft and caramel-like; flavor rich, pleasing.

Seed. Narrowly oblong; germ pore a little distal of middle; furrow narrow and shallow or closed in middle.

Sphinx

Synonym. Black Sphinx.

History. According to John McChesney, manager of the Phoenix Date Co. in 1952, Sphinx is of unknown parentage although it may be a Hayany seedling because the palms and fruit resemble those of that variety to some degree. Discovered near Phoenix, Arizona, it was transplanted to land of the Phoenix Date Co. about 1920. By 1954 there were about 12 acres of Sphinx with at least 50 palms in production. Then the land was subdivided for urban development, threatening

Figure 31. A selection from Deglet Noor seedlings, the American variety Honey, here with fruit bunches protected by paper coverings, has a medium-heavy trunk, long leaves, and fruit that are less susceptible to damage from rain and checking than Deglet Noor fruit. (Donald R. Hodel)

Figure 32. *Khalal*-stage fruit of the American variety Honey are yellow while those in *rutab* stage are amber-brown. (Donald R. Hodel)

the variety's existence. Today only one or two producing Sphinx palms exist in California; an undetermined number are bearing in Arizona.

Franklin (1929) included Sphinx on a list of the eight most important date varieties in commercial production in the Salt River Valley, at Phoenix, Arizona, in the late 1920s. Albert and Hilgeman (1935) provided a brief description along with a photograph of the fruit.

Today, there is small-scale harvesting in Phoenix on remnant trees from former groves that have been developed for residential use.

Distinguishing characters. Sphinx differs from Hayany in its slower growth; leaves with more distal curvature; shorter, less-drooping pinnae; shorter, greenish yellow fruitstalks with moderate scurf proximally; and shorter, usually slightly smaller fruit. Spines are similar, though, with a pronounced neck 3–6 cm long.

Palm. Trunk medium heavy. Leaves medium long; curvature medium increasing distally.

Leaf bases. Medium broad; green with maroon at base, mottled on younger leaves, sometimes continuous across older ones; scurf slight.

Spines. 21–25, occupying about $^1/_7$ of blade; mostly solitary, a few groups of two; 12–24 cm long; slender

Figure 33. Attractive, red *khalal*-stage fruit of the American variety Sphinx are the perfect complement to the golden yellow rachillae and peduncle. (Donald R. Hodel)

and flexible; neck 2–6 cm long, definite; rachis angles 45–90° for single, 30–60° for pairs; a-r divergence 20–30°.

Pinnae. Drooping slight; longest 67–68 cm × 1.2–2.0 cm; widest 41–47 cm × 3.8 cm; terminal 32–33 cm × 2.4–2.5 cm; valley angle 120–140° proximally to 130–145° at apex; dorsal angle 165–180° proximally to 170–175° at apex; rachis angles proximally 20–50°; apical divergence 65–100°; BSI 30–35%; in groups of two proximally, some in groups of three distally, indistinct near apex; classes definite except near apex.

Fruitstalks. Greenish yellow; moderate scurf on lower portion; short; medium heavy.

Fruit. *Khalal* color medium red ("nopal red") (Figure 33), some dull yellow ("xanthine orange") showing in background of many; *rutab* and *tamar* dark brown to nearly black; oblong-oval, greatest diameter sometimes slightly proximal or distal of middle; 41–48 mm × 26–28 mm; calyx prominent, margin rounded, entire or slightly broken; skin medium thick, blistering somewhat; flesh soft; rag slight and unobjectionable; flavor good, mild; late ripening.

Seed. Medium brown; oblong-elliptical; 25–30 mm × 8.8–9.8 mm; germ pore variable; furrow usually closed and ventral surface flattened in middle, but sometimes medium deep and medium wide.

Notes/culture. Fruit are subject to severe, irregular checking, and rain or humid weather has caused considerable spoilage although such damage is less serious than that on Hayany. Annual yields of 300 pounds or more of fruit per tree have been reported. Vertical growth of the palm is nearly as slow as that of Khadrawy. Offshoot production is prolific, with 25 to 30 per palm being common.

Sphinx is in the Thermal and Tempe germplasm collections (Appendix C) and that of the Yuma Mesa Agricultural Center at Somerton, Arizona.

Tabarzal

History. The origin of Tabarzal is uncertain. According to A. J. Shamblin, who was in charge of inspection for *Parlatoria* date scale during the period when commercial importations were being made in the early 1900s, the original palm was a seedling near Mecca in the Coachella Valley, California. S. S. M. Jennings, an early resident near Mecca, reported that the original palm came from a planting of imported offshoots elsewhere in the Coachella Valley, but there is no record of any other similar palms among the imported plantings. As of this writing 12 producing palms exist (Figure 34).

M. R. Sheets had acquired the original palm and planted it on property about 3 miles southeast of

Figure 34. Only twelve fruiting plants of the American variety Tabarzal still exist. (Donald R. Hodel)

Mecca, California. Jess Wise, who purchased the Sheets property, developed Tabarzal and trademarked the name. He selected the name because a visitor from Egypt once remarked that the date resembled a variety by that name in the Old World. Wise had 16 acres with at least 50 palms in production in 1955.

Distinguishing characters. The very large fruit and its texture, softer than that of the average soft varieties, distinguish Tabarzal from other varieties grown in the United States. Leaves are only medium long and fruiting rachillae are thicker than those of most varieties.

Palm. Trunk slender to medium heavy. Leaves medium long; curvature moderate, increasing distally.

Leaf bases. Medium broad; green, some chocolate colored areas at base of oldest; scurf slight.

Spines. Few to a moderate number; occupying about ⅕ of blade; about one-half to two-thirds of them solitary, others in groups of two but usually not closely so; 6–12 cm long; medium heavy, stiff; neck lacking; rachis angles 20–45°; a-r divergence 20–25°.

Pinnae. Drooping slight; longest 54–55 cm × 1.7–2.3 cm; widest 41–42 cm × 4.4 cm; terminal 22–23 cm

× 2.4 cm; valley angle from 95° proximally to 130° at apex; dorsal angle from 140–155° proximally to 180° at apex; apical divergence 75–35°; in groups of two at base with many groups of three beginning a little proximally, indistinct distally of midblade; classes definite.

Fruitstalks. Yellowish green; scurf inconspicuous; medium long; medium heavy.

Fruit. *Khalal* light yellow, some with pale greenish tint; *rutab* light amber brown; *tamar* reddish brown; oblong-oval to oblong-ovate; 46–54 mm × 27–29 mm; calyx moderately prominent, three-parted; skin thin, tender, blistering slightly; flesh 6–8 mm thick, soft; rag slight, tender; flavor mild, pleasing; early ripening (a little later than Khadrawy).

Seed. Medium brown; narrowly oblong, greatest diameter usually near apex, diameter near base sometimes slightly greater than just distally, deformities sometimes present; 26–30 mm × 7.3–8.4 mm; germ pore near center but variable; furrow variable, narrow and shallow, widening a little at apex, more so at base, sometimes closed in midportion.

Notes/culture. Fruit are subject to slight checking in short, transverse or somewhat irregular, apical lines. Damp weather may cause considerable spoilage. Yields purportedly equal those of Deglet Noor.

Tabarzal is in the Thermal and Tempe germplasm collections (Appendix C).

T-R

History. Also called Triumph and from the same seed lot as Empress, T-R is a Thoory seedling that E. K. Davall planted on his property near Cathedral City, California, in 1916 or 1917. Its name, T-R, refers to the third row, and probably alludes to its location in a date grove. In 1955 there were about 40 palms, but the variety had not been propagated in recent years. Presently 12 producing palms exist (Figure 35). Fruit are subject to slight checking in irregular apical lines and in faint transverse lines proximally. The palm somewhat resembles Thoory but leaves are more curved, the crown is more open, and leaf bases have heavy scurf on the edges. Fruitstalks are yellow with moderate scurf, short, and medium heavy.

Fruit. *Khalal* pale yellow with slight greenish tinge; *rutab* amber; *tamar* reddish brown; oblong, sloping from a little distal of middle to bluntly pointed apex; 43–52 mm × 19–22 mm; calyx flattened, three-parted or nearly so; skin medium thick, adhering to flesh in coarse wrinkles; flesh soft and melting; rag absent or slight and unobjectionable; flavor pleasing; early ripening.

Seed. Oblong and sometimes slightly spatulate; 23–29 mm × 7.4–8.7 mm; germ pore about ¼ seed length from apex; furrow closed or shallow and narrow to medium wide.

Figure 35. Only twelve fruiting plants of the American variety T-R still exist. (Donald R. Hodel)

T-R is in the Thermal germplasm collection (Appendix C).

NON-COMMERCIAL AMERICAN VARIETIES

Amarillo

History. A large, soft date of unknown parentage, Amarillo first attracted attention at the Date Palm Collection of Arizona State Agricultural Station at Tempe. Only a few offshoots were planted and, based on their performance at the University of Arizona's Yuma Experiment Station, they were rated as a date of poor quality because damp weather seriously damaged the fruit. The palm is suggestive of a cross between Maktoom and Saidy. *Amarillo* is the Spanish word for yellow; the name refers to the color of the fruit.

Fruit. *Khalal* yellow; *rutab* amber; *tamar* reddish brown; oblong-obovate; calyx prominent, margin moderately broken; skin medium thick; ripening midseason.

Seed. Oblong; germ pore central or slightly below; furrow narrow and medium deep.

Amber Queen

History. Amber Queen, like Haziz, is believed to have originated from several thousand seed, most of them from Deglet Noor, that J. H. Northrop planted on his property near Indio, California, between 1911 and 1917. B. H. Hayes was primarily responsible for developing this variety. Hayes acquired part of the Northrop property in 1924 and, recognizing the merit of this variety, selected and propagated offshoots until he had about four acres and at least 50 palms in production when he sold his estate to M. I. Kornell in 1955.

Distinguishing characters. The palm is similar to Haziz but differs in its shorter, more arched, slightly lighter, more glaucous leaves; more numerous spines, mostly in groups of two but usually two or three groups of three; shorter, less drooping pinnae; slightly smaller rachis angles and apical divergence; shorter fruiting rachillae; and smaller, softer fruit.

Palm. Trunk medium heavy. Leaves medium to long; curvature medium to pronounced, increasing distally.

Leaf bases. Medium broad; green, slight yellowish tint with some maroon mottling on oldest; scurf slight.

Spines. 46, occupying about $\frac{1}{3}$ of blade; about one-third of them solitary, others in groups of two except for one or two groups of three on some leaves; 8–17 cm long; medium heavy, stiff; without neck; rachis angles of paired spines 20–60°; a-r divergence 25–45°.

Pinnae. Drooping slight; longest 69–70 cm × 2.0–2.3 cm; widest 53–55 cm × 4.8–4.9 cm; terminal 32–34 cm × 3.0–3.3 cm; valley angle from 40° proximally to 100° at apex; dorsal angle from 150° proximally to 160° at apex; apical divergence 75–85°; BSI 65–70%; in groups of two, three, and four, but indistinct distally of middle because of coalescence; classes definite throughout.

Fruitstalks. Orange yellow; moderate scurf near base; medium long; medium heavy.

Fruit. *Khalal* yellow ("apricot yellow" to "antimony yellow"), *rutab* amber, *tamar* a deeper shade of amber; oblong-oval; 41–47 mm × 21–24 mm, greatest diameter a little above middle; calyx moderately prominent, margin rounded and entire or slightly broken; skin medium thick, usually adhering to flesh in coarse wrinkles, occasional blistering; flesh soft; a little rag, which sometimes may be rather tough; flavor good; early ripening.

Seed. Medium brown above germ pore, light gray and buff below; narrowly oblong-elliptical; 22–26 mm × 6.1–7.6 mm; germ pore a little below center; furrow closed in middle or shallow and narrow, opening more toward base and apex.

Notes/culture. Fruit are only slightly less subject to checking than those of Deglet Noor. The skin tends to separate from the flesh if left on the bunch too long. There has been considerable fruit spoilage in wet years.

Andrate

History. Andrate is one of many selections of putative "Medjool" seedlings that Francis Heiny planted at his garden in Brawley, California, about 1911 (these seedlings must have originated from palms derived from imported seed, perhaps from Medjool but certainly not from offshoots of true Medjool, as Medjool had not yet been introduced). Heiny had 12 young palms of this variety in 1955. Fruit are about the same size as those of Medjool but are more angular and of lesser quality. The fruit appear to be fairly rain tolerant, not checking or souring from inclement weather. The palm resembles Medjool but has shorter and slenderer fruitstalks.

Fruit. *Khalal* yellow; *rutab* reddish brown; *tamar* dark or blackish brown; oblong; 43–54 mm × 25–27 mm; perianth set in slight depression; calyx moderately elevated, three-cleft; skin adhering to flesh in fine wrinkles; flesh soft; rag slight to moderate but often rather tough; flavor fair; late ripening.

Seed. Oblong-elliptical, sometimes slightly spatulate, wings usually present; 23–27 mm × 9.4–10.3 mm; germ pore central or slightly below; furrow medium in width and depth.

Black Lucifer

History. Black Lucifer is another of the many selections of putative "Medjool" seedlings that Francis Heiny planted at his garden in Brawley, California, prior to 1955. This variety is a heavy producer, according to Heiny. Fruit have very little tannin in the *khalal* stage, are subject to considerable shrinkage, and do not cure well. Checking occurs in longitudinal lines.

Fruit. *Khalal* dark red; *rutab* and *tamar* nearly black with moderate to heavy bloom lending a purplish cast; oblong-oval; 42–46 mm × 24–28 mm; perianth in slight depression; calyx yellow, slightly to moderately elevated, three-parted or nearly so; skin adhering to flesh in moderate wrinkles; flesh soft; rag slight; flavor fair; ripening midseason.

Seed. Oblong and slightly spatulate-elliptical; 26–29 mm × 11.0–12.6 mm; germ pore central; furrow medium to broad, medium deep; ridges present but not pronounced.

Black Medjool

History. Black Medjool is one of the many selections of putative "Medjool" seedlings that Francis Heiny planted at his garden in Brawley, California, about 1911. Heiny had five palms of this variety in 1955 and there were about 25 others at Sunnipalms, a commercial date garden near Indio, California. Fruit have no resemblance to those of Medjool. They are not very susceptible to checking and are undamaged by occasional rains. The palm differs from Medjool in its slightly more curved leaves; longer and narrower pinnae, the more distal ones with a wider rachis angle; and fewer spines. Fruitstalks are orange-yellow with slight to medium scurf, medium long, and medium heavy.

Fruit. *Khalal* deep red; *rutab* and *tamar* nearly black, moderate bloom lending a purplish cast; oblong-oval; 37–41 mm × 21–23 mm; calyx yellow (margin tinged with red), moderately prominent, margin slightly broken to nearly three-parted; rag slight, unobjectionable; flavor very sweet, somewhat cloying; early ripening.

Seed. Oblong-elliptical, often slightly spatulate; 22–29 mm × 8.9–10.2 mm; germ pore usually slightly below middle; furrow closed or narrow and shallow, sometimes pitted apically.

Black Royal

History. A seedling of unknown parentage, C. W. Hyde selected and propagated Black Royal at his garden near Coachella, California. In 1955 R. B. Van Blaricom owned Hyde's property and had 40 palms of this variety at the site. Fruit are very susceptible to damage from rain and high humidity. They check badly in short transverse lines from middle to apex, tips shrivel and darken, and heavy losses occur from rot and drop. Black Royal is distinguished by its compact crown of leaves; wide pinnae with wide rachis angles; widely separated proximal pinnae groups; and bands of fiber ³⁄₄ to 1¹⁄₄ inches wide on the trunk. Fruitstalks are greenish yellow, without scurf, short to medium long, and medium heavy.

Fruit. *Khalal* light yellow; *rutab* dull brown; *tamar* nearly black but retaining a deep dull brown apically; oblong, slightly obovate; 44–52 mm × 21–24 mm; skin medium thick; flesh soft; rag slight, unnoticeable; flavor pleasing; ripening midseason.

Seed. Narrowly oblong or oblong-elliptical; germ pore a little above middle; furrow narrow to medium wide and medium deep, sometimes widening at apex with a peculiar spatulate exposure of tissue.

Deglar

History. Reflecting the name of its parents, Deglar is a cross using a seedling of Deglet Noor as the pollen plant and Rhars as the seed plant. Francis Heiny had made the cross and selected the seedling at his garden at Brawley, California. Heiny had four palms of this variety in 1955. Fruit are subject to considerable transverse checking.

Fruit. *Khalal* yellow; *rutab* amber; *tamar* reddish brown; oblong-oval; 42–44 mm × 21–23 mm; calyx prominent or moderately prominent, margin moderately broken to three-cleft; skin thin; flesh soft; rag slight; flavor rich and pleasing; early ripening.

Seed. Elliptical; 21–24 mm × 8.2–9.1 mm; germ pore central; furrow narrow and shallow.

Desert Dew

History. Brooks and Olmo (1950) reported that C. D. McGinnis selected Desert Dew from offshoots taken from Bard Valley, California, to the McGinnis date grove in Yuma, Arizona, in 1909. It was introduced commercially in 1946. Nixon observed this variety in 1951, 1952, and 1954 at the McGinnis place, which had since been sold. He noted that the palms were uncared for and not fruiting, most being so badly crowded and long neglected in overgrown nursery rows that it was difficult to determine their identity. About 100 offshoots were said to have been planted near Yuma by the 1950s, but were not bearing at the time. Although reports indicated that fruit did not crack or check, Nixon observed a slight amount of transverse and apical checking, as is typical of Deglet Noor, in 1951, a year when very little checking occurred on dates, even on susceptible varieties. The palm somewhat resembles that of Deglet Noor. Fruitstalks are greenish yellow with slight to medium scurf proximally, medium long, and medium heavy.

Fruit. *Khalal* red (near "nopal red" or "eugenia red"); *rutab* dull brown; *tamar* a deeper, somewhat purplish brown; oblong with rounded base and apex; 39–46 mm × 18–21 mm; calyx yellow, moderately prominent, margin rounded and entire or slightly broken, corolla pink; skin thin, tender, usually adhering to flesh; flesh very soft and melting; rag absent or slight and unobjectionable; flavor rich and sweet but cloying; ripening midseason.

Seed. Narrowly oblong or oblong-elliptical; 23–27 mm × 7.1–8.3 mm; germ pore near middle but variable; furrow narrow and shallow.

El Toby

History. El Toby originated from five very old palms that Mulford Winsor had discovered on abandoned farmland in the Yuma Valley of Arizona in 1923. Their previous history is unknown. From the five original palms, Winsor had propagated 100 plants on about 2 acres in his date garden on the Yuma Mesa about 10 miles south of Yuma before it was sold to W. G. Long in 1954. By 1955 there were at least 50 palms in production.

Distinguishing characters. From a distance, the crown of El Toby resembles that of Zahidi, but is somewhat less compact. El Toby also has less glaucous leaf bases with a solid maroon band near the fibers. The small- to medium-sized fruit are bright red before ripening and later become nearly black.

Palm. Trunk medium to heavy. Leaves long; curvature slight, increasing distally.

Leaf bases. Medium wide; maroon across base with some slight brown portions near the fiber, glaucous green above; scurf slight.

Spines. 26, occupying about ⅕ of blade; in groups of two in distal half of spine area, solitary proximally; 6–16 cm long; slender but stiff; neck 1–2 cm long, indefinite; rachis angles of paired spines 10–15°; a-r divergence 5–15°.

Pinnae. Drooping slight, a few moderately so near tip of blade whereas those proximally usually stiff; longest 61 cm × 3.2 cm (near tip of blade); widest 48 cm × 4.2 cm; terminal 41 × 3.0 cm; valley angle from 75° proximally to 120° at apex; dorsal angle from 210° at base to 180° at apex; rachis angles proximally 15–40°; apical divergence 50°; BSI 35–40%; in groups of two and three and a few fours, mostly definite except near apex; classes mostly definite.

Fruitstalks. Orange yellow; slight to moderate scurf near base; medium to long; slender.

Fruit. *Khalal* bright red (near "carmine"); *rutab* and *tamar* nearly black with moderate bloom; ovate or oblong-oval; 30–40 mm × 17–19 mm; calyx yellow, moderately prominent, margin rounded and entire or slightly broken to nearly three-cleft; skin medium thick, rather tough, tending to blister; flesh soft; rag slight, unobjectionable; flavor rather strong; ripening midseason.

Seed. Medium brown; narrowly oblong; 19–22 mm × 6.5–7.2 mm; germ pore slightly proximal of middle; furrow narrow and medium deep.

Notes/culture. El Toby is slow to come into bearing and makes slow vertical growth. Fruit appear resistant to damage from rain, and only slight longitudinal checking at the base with some irregular checking apically has been observed.

Ford Date

History. Named for the person who had bought the original plant, Ford Date is a seedling of unknown parentage that Francis Heiny planted at his garden in Brawley, California. Heiny had four palms of this variety in 1955. Fruit resemble those of Deglet Noor in size, shape, and texture but are poorer in quality, a little wider at the base, and very early ripening, and are said to be less susceptible to checking.

Fruit. *Khalal* yellow with faint pinkish flush; *rutab* and *tamar* dull brown; calyx moderately prominent, margin moderately broken to three-cleft.

Seed. Oblong; germ pore slightly above middle; furrow closed or shallow and narrow.

Frau Damke

History. Frau Damke is a large, mediocre, soft date that originated from seed of unknown parentage on the former D. E. Magill ranch near Mecca, California. Max Zimmer propagated a few offshoots on his ranch near Mecca, but in 1955 there were only three surviving palms of this variety. Fruit are notable because of their large size and the unusually pronounced depression in which the very small perianth is set.

Fruit. *Khalal* yellow; *rutab* dull greenish amber; *tamar* reddish brown; oblong-ovate, sloping to somewhat bluntly pointed apex; skin medium thick; rag moderate, sometimes rather tough; ripening midseason.

Haziz

History. Haziz, like Amber Queen, is believed to have originated from several thousand seed, mostly of Deglet Noor, that J. H. Northrop planted on his property about 3 miles west of Indio, California, between 1911 and 1917. B. H. Hayes, for whom the variety was named, was primarily responsible for developing this variety. Hayes acquired part of the Northrop property in 1924 and, recognizing the merit of this variety, selected and propagated offshoots until there were approximately 14 acres with at least 50 palms in production when he sold his estate to M. I. Kornell in 1955.

Distinguishing characters. Compared to Deglet Noor, Haziz has slightly fewer spines, wider pinnae with somewhat larger rachis angles, and only medium length fruitstalks. Fruit are somewhat similar to but larger than those of Halawy and, like Halawy fruit, they sometimes have dry portions at the base.

Palm. Trunk medium heavy. Leaves long; curvature slight to medium, increasing distally.

Leaf bases. Medium broad; green with only an occasional trace of maroon mottling; scurf slight.

Spines. 37, occupying about $\frac{1}{4}$ of blade, mostly in groups of two; 10–18 cm long; medium heavy, stiff; neck 1–2 cm long, indefinite; rachis angles of paired spines 20–35°; a-r divergence 30–35°.

Pinnae. Drooping slight to moderate, irregular; longest 88–91 cm × 2.1–2.3 cm; widest 70 cm × 4.9–5.2 cm; terminal 55–57 cm × 2.9–3.3 cm; valley angle from 45° proximally to 130° at apex; dorsal angle from 145° proximally to 180° at apex; apical divergence 90°; BSI 55–70%; in groups of two, three, and a few fours, indistinct distally of middle; classes definite proximally, many indefinite distally.

Fruitstalks. Orange yellow; slight traces of scurf proximally; medium long; medium heavy.

Fruit. *Khalal* yellow ("apricot yellow" to "antimony yellow"); *rutab* amber; *tamar* reddish brown ("morocco red" or "claret brown"); oblong-ovate; 44–48 mm × 22–25 mm; calyx prominent, more or less abruptly elevated, margin moderately broken to nearly three-cleft; skin medium thick, usually adhering to flesh in coarse wrinkles; flesh 5–6 mm thick, soft and melting; rag absent or slight and unobjectionable; flavor mild and pleasing; ripening midseason.

Seed. Dark brown distally of germ pore, light buff proximally; narrowly oblong-elliptical, greatest diameter usually above middle; 24–30 mm × 7.2–8.2 mm; germ pore central or a little distally; furrow shallow and medium depth or closed in midportion, opening or narrowly pitted at apex.

Notes/culture. Fruit of Haziz are only slightly less subject to rain damage than those of Deglet Noor and are similar in their susceptibility to checking. Haziz flowers one to two weeks earlier than Amber Queen and most other varieties, and the lower temperatures often prevailing then may be responsible for occasionally poor fruit set. Fruit sometimes tend to shrivel. Haziz is in the Thermal and Tempe germplasm collections (Appendix C).

Heiny

History. Heiny is one of the many selections of putative "Medjool" seedlings that Francis Heiny planted at his garden in Brawley, California, about 1911. Heiny had 20 palms of his namesake variety in 1955. Compared to fruit of Medjool, those of Heiny are similar in flavor and in their firm-soft flesh, but are slightly larger, more oblong or blocky, slightly lighter in color, somewhat more subject to spoilage following rain, and have checking in short longitudinal lines. The palm bears considerable resemblance to Medjool but differs in its stiffer, close-set pinnae with smaller rachis angles. Leaf bases have slight to moderate scurf that sometimes extends along the rachis into the proximal pinnae where it may be rather heavy. Fruitstalks are orange-yellow with slight to moderate scurf, short to medium long, and heavy.

Fruit. *Khalal* yellow; *rutab* amber; *tamar* dull reddish brown; broadly oblong and slightly obovate; 47–54 mm × 27–31 mm; perianth set in slight depression; calyx slightly to moderately prominent; skin medium thick, adhering to flesh in fine wrinkles with little blistering; flesh firm-soft; rag slight; flavor mildly rich, pleasing; early ripening.

Seed. Oblong; 27–30 mm × 10–11 mm; germ pore central or slightly above; furrow closed; ridges usually present but not pronounced.

Heinyana

History. Heinyana is another of the many selections of putative "Medjool" seedlings that Francis Heiny planted at his garden in Brawley, California, about 1911. Thirty palms of Heinyana existed in 1955, of which 10 were in the Heiny garden, 12 on a neighboring property, and the remainder in a planting 2 miles west of Indio, California. Fruit are smaller and somewhat lighter in color than those of Medjool and have about the same tolerance to rain. The palm is somewhat suggestive of Medjool but differs in the dark maroon areas on the edges of old leaf bases. Fruitstalks are orange-yellow with slight to medium scurf, medium long, and medium heavy.

Fruit. *Khalal* light yellow (perianth deeper yellow); *rutab* light amber; *tamar* light reddish brown; oblong-obovate, flattened at apex; 30–42 mm × 21–23 mm; calyx slightly to moderately prominent, margin slightly broken; skin thin, mostly adhering to flesh; flesh soft, 5–6 mm thick; rag slight, unobjectionable; flavor mild and pleasing; early ripening.

Seed. Oblong, apex rounded, base slightly pointed, wings characteristic; 20–22 mm × 6–7 mm; germ pore about ⅓ seed length from base; furrow closed or narrow and shallow.

Humbarger

History. A Deglet Noor seedling that originated at the USDA Date and Citrus Station at Indio, California, this variety was named for Oliver Humbarger, a U.S. Bureau of Indian Affairs agent on the Martinez Reservation in Coachella Valley, California, who had taken a fancy to the date and planted several offshoots in the variety collection established there in the 1920s. By 1955 T. R. Brown in the lower end of Coachella Valley had the only known palm producing Humbarger. Fruit are somewhat similar in size and shape to those of Deglet Noor but are softer and different in color.

Fruit. *Khalal* deep red; *rutab* and *tamar* nearly black; oblong-elliptical, greatest diameter slightly distal of middle; calyx moderately prominent, margin rounded and slightly broken; skin thin, shrinking with flesh in irregular folds and wrinkles; flesh soft and melting; rag slight, unobjectionable; flavor sweet and pleasing; late ripening.

Seed. Narrowly oblong; germ pore distal of middle; furrow narrow and shallow, fairly uniform.

Laguna

History. Of unknown origin and perhaps an imported variety, Laguna was purportedly one of a collection of date palms that the U.S. Bureau of Reclamation had obtained from the old Arizona Agricultural Experiment Station near Yuma and had planted at Laguna Dam a few miles up the Colorado River, hence the varietal name. F. W. Creswell secured two offshoots of Laguna and propagated from those until, in 1955, he had 16 palms in production on his ranch in the Yuma Valley. He had supplied about 75 offshoots of this variety for planting on another ranch.

Palm. Trunk heavy. Leaves stout; curvature slight to moderate, increasing distally.

Leaf bases. Wide; yellowish green with a little pale brown near the fiber line and some slight maroon mot-tling adaxially; slight to moderate scurf on edges.

Spines. Medium in number, occupying about ⅕ of blade; in groups of two; to 12–15 cm long; slender, medium stiff; neck 2–3 cm long, indefinite.

Pinnae. Medium in length and width with slight to moderate drooping; mostly in groups of two with a few groups of three midblade; classes indistinct distally.

Fruitstalks. Greenish yellow; moderate scurf proximally; medium long; heavy.

Fruit. *Khalal* yellow; *rutab* amber; *tamar* reddish brown; broadly ovate or oblong-ovate, base abrupt, apex bluntly pointed; 40–45 mm × 25–27 mm; calyx moderately prominent, margin rounded and entire or slightly broken; skin thin and tender; flesh soft; rag slight, unobjectionable; flavor pleasing; ripening mid-season.

Seed. Oblong with abrupt base and bluntly pointed apex; 20–22 mm × 9.0–9.4 mm; germ pore near or slightly distal of center; furrow medium to broad, medium deep with depression continuing from base to about one-quarter up dorsal side.

Lindy

History. Lindy is another of the many selections of putative "Medjool" seedlings that Francis Heiny planted at his garden in Brawley, California, about 1911. Heiny had about 100 Lindy offshoots and at least 50 palms in production in 1955.

Distinguishing characters. Similar to and not easily distinguished from Medjool, Lindy differs in its shorter fruitstalks and longer pinnae.

Palm. Trunk medium heavy. Leaves medium to long; curvature slight to medium, increasing distally.

Leaf bases. Medium wide; glaucous green, slight trace of maroon mottling on oldest leaf bases; scurf slight.

Spines. 38, occupying about ⅕ of blade; about half of them closely spaced near the base, others in groups of two and usually one or two groups of three; 14–20 cm long; medium to heavy, stiff; neck lacking; rachis angles of paired spines 15–25°; a-r divergence 10–15°.

Pinnae. Drooping moderate; longest 100 cm × 2.6 cm; widest 55 cm × 5 cm; terminal 25 cm × 2 cm; valley angel from 70° proximally to 140° at apex; dorsal angle from 170° proximally to 180° at apex; rachis angle proximally 10–25° for antrorse and retrorse; apical divergence 75°; BSI 40–50%; in groups of two and three, some fours midblade, somewhat indistinct near apex; classes definite throughout.

Fruitstalks. Orange yellow; scurf slight to medium proximally; very short; heavy.

Fruit. *Khalal* pink (near "ferruginous"); *rutab* amber brown; *tamar* dull reddish brown with bloom lending a purplish cast; oblong; 35–50 mm × 26–32 mm; calyx moderately prominent, margin rounded to moderately broken; skin medium thick, adhering to flesh in coarse wrinkles; flesh soft but firm; rag slight, unobjectionable; flavor very similar to Medjool; early ripening.

Seed. Medium brown; oblong or oblong-elliptical, usually with wings; 20–23 mm × 6–10 mm; germ pore near middle but variable; furrow shallow and narrow to medium wide, or closed, sometimes pitted at apical end (wings cause variability).

Notes/culture. Lindy is very similar to Medjool in both fruit and palm characters. Fruitstalks of Lindy, however, are even shorter than those of Medjool; indeed, a large percentage of them are too short for proper handling of the bunch, their small size being a serious drawback to this variety as a commercial date.

Long Yellow

History. Long Yellow is another of the many selections of putative "Medjool" seedlings that Francis Heiny planted at his garden in Brawley, California, about 1911. Heiny had only two palms in 1955 but he reported that another grower had planted 15 offshoots at Niland, California. Fruit have no resemblance to those of Medjool and they are subject to severe checking in long, transverse lines, but do not easily sour.

Fruit. *Khalal* light pink; *rutab* amber; *tamar* reddish brown; oblong-obovate; 40–42 mm × 19–21 mm; calyx prominent, margin rounded-triangular, usually slightly broken; skin thin, mostly shrinking with flesh in wrinkles; flesh soft; rag light; flavor good; ripening midseason.

Seed. Oblong-elliptical, often slightly spatulate; 21–22 mm × 8.2–9.0 mm; germ pore usually slightly proximal of center; furrow narrow and shallow, opening more at base and apex.

Maktoom No. 1

History. This variety is a Maktoom seedling that D. H. Mitchell propagated at the Coachella Valley Fruit Co. near Coachella, California. In 1955 there were seven palms of this variety in production while 15 offshoots had recently been planted and 30 more had been sold for planting in the Coachella Valley. Fruit are very similar to those of Maktoom, but are reported to set poorly and ripen early. The palm suggests a cross between Khadrawy and Zahidi. Fruitstalks are orange-yellow with moderate scurf, medium long, and slender to medium heavy.

Fruit. *Khalal* yellow with a slight pink cast; *rutab* amber; *tamar* dull reddish brown; oblong-oval; medium size; calyx moderately prominent; skin medium thick; flesh soft; rag slight; flavor good; ripening midseason.

Seed. Dark brown, narrowly oblong; 33.0 mm × 6.6 mm; germ pore central or nearly so; furrow closed or narrow and shallow.

Malakai

History. Of unknown origin, Malakai is believed to be a seedling from the Carl Shepard ranch about 3 miles south of Indio, California. There were seven Malakai palms in production in 1955.

Fruit. *Khalal* red; *rutab* reddish black; *tamar* nearly black, the perianth remaining yellow with reddish tints on edge; oblong-oval; 42–45 mm × 19–21 mm; calyx moderately prominent, margin rounded and slightly broken; skin medium thick, blistering considerably; flesh soft and melting; rag slight; flavor good; early ripening.

Seed. Medium brown; narrowly oblong; 25–32 mm × 7–8 mm; germ pore variable; furrow variable.

Metzler's Black

History. Sometimes called Metzler's Patent Leather Black Date, this variety is a seedling of unknown parentage that V. E. Metzler selected and propagated on his former ranch near Coachella, California. The varietal name reflects the appearance of its ripened black fruit.

Fruit. *Khalal* dark red; *rutab* and *tamar* black, rather glossy, especially after moderate bloom is removed by cleaning; oblong to oblong-ovate; 43–50 mm × 22–28 mm; calyx moderately prominent, margin rounded and slightly broken; skin rather tough, tending to separate from flesh in curing; flesh soft; rag slight to moderate, occasionally a little tough; flavor mild when fresh, becoming rather heavily sweet after curing; ripening midseason.

Seed. Oblong-elliptical; 24–27 mm × 7.5–9.2 mm; germ pore near center, furrow variable.

Peggy Ann

History. Peggy Ann is another of the many selections of putative "Medjool" seedlings planted by Francis Heiny at his garden in Brawley, California, about 1911. L. W. Eddins propagated this variety on his ranch near Calipatria, California, and named it for his daughter. In 1955 there were 16 bearing palms and a number of recently planted offshoots.

Palm. Similar to Medjool. Leaves with moderate curvature increasing distally.

Leaf bases. Medium to wide; only slight traces of scurf.

Spines. Numerous; in groups of two with a few groups of three; broad and stiff; neck lacking; rachis angles 30–60°; a-r divergence about 45°.

Pinnae. Drooping slight to moderate on lower leaves; medium in length and width, terminal a little less than 1/2 length of longest; apical divergence 80–100°; BSI 40–60%.

Fruitstalks. Orange-yellow; scurf lacking; medium long, medium heavy.

Fruit. *Khalal* pale, slightly greenish yellow; *rutab* dull amber; *tamar* reddish brown, some specimens occasionally retaining small, dry, buff colored areas at base; oblong, slightly ovate; 46–50 mm × 25–28 mm; calyx moderately prominent, margin slightly broken to three-parted; skin thin in rather coarse wrinkles; flesh soft; rag lacking or unnoticeable; flavor mildly rich and pleasing; ripening midseason.

Seed. Light grayish brown; elliptical, some slightly spatulate; 26–31 mm × 10–11 mm; germ pore variable; furrow narrow to broad, shallow to medium deep.

Notes/culture. Fruit are similar to those of Medjool but are narrower with slight, irregular, apical checking. Peggy Ann is in the Tempe germplasm collection (Appendix C).

Plum Date

History. A seedling of unknown parentage, Mrs. C. E. Cast selected and propagated this variety in her Garden of the Setting Sun at Mecca, California, and had seven palms in 1955. The palm resembles Khadrawy but differs in its more curved leaves and longer fruit-stalks. It also lacks the slightly longer terminal pinnae that give a somewhat "fishtail" appearance to the leaf tip, a distinctive character of Khadrawy.

Fruit. *Khalal* yellow; *rutab* light brown; *tamar* amber; oblong, or oblong-oval; 40–48 mm × 26–28 mm; perianth in slight depression; calyx flattened, margin rounded and entire or very slightly broken; skin rather tough, shrinking with flesh in coarse wrinkles and folds, some blistering; flesh soft; rag slight, unobjectionable; flavor mild and pleasant; ripening midseason.

Seed. Elliptical, often slightly spatulate; 24–27 mm × 8.5–9.5 mm; germ pore central, or slightly above; furrow closed in middle opening a little toward base and apex.

Precioso

History. A seedling of unknown parentage, Mrs. C. E. Cast selected and propagated Precioso in her Garden of the Setting Sun at Mecca, California, and had about 12 producing palms in 1955. Fruit are subject to occasional apical checking in irregular lines, some being long with rather deep breaks. The palm resembles Khadrawy but differs in its longer leaves and pinnae and more numerous and longer spines. Fruitstalks are greenish yellow with sparse or no scurf, short, and medium heavy.

Fruit. *Khalal* light yellow; *rutab* brown; *tamar* amber to reddish brown; oblong, slightly obovate; 46–52 mm × 22–25 mm; calyx moderately prominent, margin rounded and slightly broken to almost three-parted; skin thin and tender; flesh soft and melting; rag slight

and unobjectionable; flavor rich and sweet but somewhat cloying; ripening midseason.

Seed. Narrowly oblong, slightly spatulate; 26–32 mm × 7.2–8.9 mm; germ pore variable but near center; furrow variable.

Rubidor

History. Of unknown parentage, Rubidor was growing on the south side of Broadway about ³⁄₈ mile east of Arizona Road in Mesa, Arizona, when W. J. Van Spankeren acquired the property. By 1955 he had 28 large palms in full production and 26 small ones just beginning to fruit. Van Spankeren formed the varietal named by combining the names of his two daughters.

Distinguishing characters. Wide rachis angles are characteristic of this variety.

Palm. Leaves medium long; curvature moderate, increasing distally.

Leaf bases. Medium wide; green with ochraceous yellow portions proximally and occasional traces of maroon on the edges; scurf lacking or slight.

Spines. 30–34, occupying about ¹⁄₅ of blade; about half of them in groups of two; to 12–16 cm long; medium heavy.

Pinnae. Drooping slight; rather short, medium wide; wide rachis angles; mostly in groups of two and a few groups of three, distinct except near apex.

Fruitstalks. Greenish orange; lacking scurf; medium long; slender.

Fruit. *Khalal* light red with ochraceous portions on some; *rutab* dark reddish brown; *tamar* a deeper shade, nearly black; rounded-obovate; 30–37 mm × 22–24 mm; calyx moderately prominent, margin slightly to moderately broken; skin rather tough, blistering considerably; flesh soft; rag slight, unobjectionable; flavor heavily sweet and quickly cloying.

Seed. Medium brown; oblong with greatest diameter near base; 20–24 mm × 6.3–8.0 mm; germ pore variable; furrow closed with ventral surface flattened with shallow, medium-broad opening at apical end.

Shields 6–7

History. This variety is believed to have originated from Deglet Noor seed that J. H. Northrop planted on his property near Indio. B. F. Shields later acquired Northrop's property, selected and propagated the variety, and in 1955 had 50 palms, some half of them recently planted offshoots. Pinnae are long and broad and appear to open more widely than in most varieties. Fruitstalks are yellowish green with slight to moderate scurf along proximal edges, long, and slender.

Fruit. *Khalal* yellow; *rutab* amber; *tamar* reddish brown; oblong-oval; 36–41 mm × 21–23 mm; calyx moderately prominent, margin entire and rounded or slightly broken; skin thin, blistering slightly; flesh soft and melting; rag slight, unobjectionable; flavor good; ripening midseason.

Seed. Oblong-elliptical; 21–24 mm $\times$ 6.7–7.5 mm; germ pore about $\frac{1}{3}$ seed length from apex; furrow closed in middle or shallow and narrow.

Smoky

History. Smoky is another of the many selections of putative "Medjool" seedlings planted by Francis Heiny at his garden in Brawley, California, about 1911. Heiny had seven palms of this variety in 1955. The varietal name is based on the fruit, which have a moderate bloom lending a "smoky" appearance. Although little fruit spoilage has been observed, there is sometimes considerable checking in longitudinal lines from base to apex. The palm has a slender trunk, leaves with slight curvature, stiff pinnae with rather wide rachis angles, and short fruitstalks.

Fruit. *Khalal* dull yellow or greenish yellow; *rutab* and *tamar* dull, dark brown with a moderate bloom lending a "smoky" appearance; broadly oblong, perianth and stigmatic scar both in slight depression; 43–46 mm $\times$ 29–31 mm; calyx moderately prominent, margin rounded and usually slightly broken; skin medium thick, mostly adhering to flesh in fine wrinkles; flesh firm; considerable rag; flavor mild and pleasing; very late ripening.

Seed. Oblong-oval or slightly spatulate; 24–26 mm $\times$ 10–11 mm; germ pore somewhat variable but near center; furrow medium wide and shallow to medium deep.

Sultany

History. Sultany is a Deglet Noor seedling that Francis Heiny planted at his garden in Brawley, California. A. J. Homla later purchased and propagated it on his ranch near the border with Mexico about 6 miles east of Calexico, California. In 1955 E. A. Bruce owned the property and had 50 palms in production.

Distinguishing characters. The palm is somewhat suggestive of Deglet Noor but differs in a number of respects: Sultany leaves have less of an olive-green color, its offshoots appear as far as 5 feet up the trunk, its *khalal* fruit are light yellow, and it has soft ripe fruit.

Palm. Trunk medium heavy. Leaves medium long; curvature slight, evenly distributed.

Leaf bases. Medium wide; green, slightly glaucous adaxially, yellowish near fibers; scurf slight.

Spines. Numerous; occupying about $\frac{1}{4}$ of blade; in groups of two, sometimes with one or two groups of three; 5–20 cm long; medium broad, medium to stiff; neck 2–3 cm long, indefinite; rachis angles of paired spines 20–55°; a-r divergence 25–55°.

Pinnae. Drooping slight to moderate; longest 72 cm $\times$ 1.7 cm; widest 60 cm $\times$ 3.5 cm; terminal 37 cm $\times$ 2.2 cm; valley angle from 100° proximally to 120° at apex; dorsal angle from 165° proximally to 180° at apex; rachis angles proximally 20–35°; apical divergence 65°; BSI 40–45%; in groups of two and three, and occasionally four distal of middle, distinct except near apex; classes definite.

Fruitstalks. Greenish yellow; scurf slight; medium long; medium heavy.

Fruit. *Khalal* light yellow; *rutab* dull amber; *tamar* reddish brown; oblong-oval; 37–42 mm $\times$ 21–23 mm; calyx moderately prominent, margin rounded and slightly broken to almost three-parted; skin tender; flesh soft, sometimes a thin, brown tough layer just beneath skin near base; flavor very sweet, somewhat cloying; late ripening.

Seed. Light grayish brown; oblong-elliptical; 20–22 mm $\times$ 7.8–8.3 mm; germ pore near middle but somewhat variable; furrow closed in middle and slightly pitted at apex, or may be shallow and narrow distal of middle.

Notes/culture. Fruit are subject to severe checking, which occurs in irregular but mostly transverse lines from middle to apex with darkening of the skin. Damp weather may cause considerable rot and drop. Sultany is also the name of a recognized Egyptian variety (Mason 1915b) unknown in the United States. How the name came to be applied to an American variety is unknown.

Sweetheart

History. Sweetheart is another of the many selections of putative "Medjool" seedlings that Francis Heiny planted at his garden in Brawley, California. Heiny had three palms of this variety in 1955. He had sold a few offshoots, but the number surviving was unknown. Although the fruit are large, they are still smaller, softer, darker, and of poorer quality than those of Medjool.

Fruit. *Khalal* yellow with slight pinkish cast; *rutab* and *tamar* reddish brown, nearly black at base; oblong-ovate; perianth set in moderate depression; calyx moderately prominent, margin three-cleft; skin thin and tender, adhering to flesh, shrinking in coarse wrinkles; early ripening.

Seed. Oblong-elliptical, usually with one or more wings or ridges; germ pore variable; furrow variable.

Wieman's Date

History. A Medjool seedling, six palms of this variety were on the Wieman Ranch 5 miles west of Brawley, California, in 1955. The palm closely resembles Medjool. Nixon noted that the palms had not been hand-pollinated when he observed them, and that the few fruit were wind-pollinated and fairly large, like those of Medjool but nearly round. Fruit are said to be early ripening.

Wonderful Black

History. A seedling of unknown parentage, Mrs. C. E. Cast selected and propagated Wonderful Black in her Garden of the Setting Sun at Mecca, California, and had eight palms in 1955. Stiff leaves with short, stiff pinnae, a long spine area, and long, slender fruitstalks distinguish this variety.

Fruit. *Khalal* deep red; *rutab* and *tamar* nearly black with moderate bloom; ovate to oblong-oval; 41–49 mm × 23–26 mm; calyx moderately prominent, margin slightly to moderately broken, yellow with red on edges or entirely red; skin rather thick, cures with coarse wrinkling and some blistering; flesh soft; rag slight, unobjectionable; flavor rich, somewhat cloying; ripening midseason.

Seed. Elliptical; 20–26 mm × 7.8–8.7 mm; germ pore central or slightly distal; furrow closed in middle, or shallow and medium wide, pitted at apical end.

ADDENDUM TO AMERICAN VARIETIES

Three additional varieties of American origin that have recently been developed or selected are formally described below. Little is known of their specific cultural requirements and fruit handling and storage properties, but all three show promise and are worthy of additional evaluation.

Benji

History. Developed by long-time Coachella Valley date grower Ben Laflin, Jr., of Laflin Date Gardens (now Oasis Date Gardens), Benji is the product of a pistillate Barhee crossed with a fourth-generation-backcrossed staminate Medjool. The staminate Medjool was a product of the breeding program at the USDA Date and Citrus Station in Indio, California. Laflin obtained the staminate Medjool when the station closed in 1980, and Benji first fruited in the late 1980s.

A dark amber date of considerable merit, Benji is outstanding for its soft, smooth, melting flesh of buttery texture and flavor. Unfortunately, it is not yet commercially available and Laflin has only one palm. Fruit of Benji have shown little or no checking, but its susceptibility to other damage from rain or humid weather is unknown. The palm's appearance is intermediate between the two parents varieties, Barhee and Medjool. The Medjool parentage is clearly evident, though, in its "winged" seed. Like Barhee, Benji produces very few offshoots.

Fruit. *Khalal* yellow; *rutab* amber; *tamar* dark brown with glaucous bloom; broadly ovate to rounded; 30–37 mm × 27–32 mm; skin thin, shrinking with flesh in irregular folds or blistering somewhat; flesh thick, soft, smooth, translucent, seldom with more than a trace of rag; flavor rich, delicate, exceptionally pleasing with a hint of butter; ripening midseason.

Seed. Brownish mottled with dark brown and gray; broadly oblong-elliptic, broadest slightly above middle, and there with one or more distinctive wings or ridges, apex broadly rounded and slightly mucronate, base narrowly rounded; 18–20 mm × 8–10 mm; germ pore central; furrow more or less closed.

8-Ball

History. Also developed by Ben Laflin, Jr., 8-Ball, like Benji, is the product of a pistillate Barhee crossed with a fourth-generation-backcrossed staminate Medjool. The staminate Medjool was a product of the breeding program at the USDA Date and Citrus Station in Indio, California. Laflin obtained the staminate Medjool when the station closed in 1980, and 8-Ball first fruited in the 1990s.

A very dark, nearly black date of some merit, 8-Ball is attractive in appearance and noted for its soft, smooth flesh and pleasing flavor. Laflin considers it one of the best black dates, and it "cures down" better than other black dates. Not yet widely available, Oasis Date Gardens has only one producing tree and markets a limited quantity of fruit. Fruit of 8-Ball are subject to only slight checking at the apex but its susceptibility to other damage from rain or humid weather is unknown. The palm is intermediate in appearance between the two parent varieties. Unlike Benji, 8-Ball lacks characteristic wings or ridges on the seed, belying its Medjool parentage. Although it produces significantly more offshoots than Benji, many offshoots of 8-Ball are high on the trunk.

Fruit. *Khalal* red; *rutab* and *tamar* very dark brown, nearly black, with glaucous bloom lending a somewhat purplish cast; broadly ovate; 33–40 mm × 26–30 mm; skin medium, shrinking with and adhering to flesh in irregular folds; flesh moderately soft, smooth with little rag; flavor pleasing; ripening midseason.

Seed. Brownish with some areas of gray; oblong-elliptic, broadest at or slightly distal of middle, apex rounded and prominently mucronate, base pointed; 16–24 mm × 8–10 mm; germ pore central; furrow wide, shallow.

Mariana

History. Coachella Valley date grower Doug Adair of Pato's Dream Date purchased several offshoots, purportedly of the Halawy variety, in the late 1980s. The

Figure 36. *Rutab*-stage fruit of the American variety Mariana are amber. (Donald R. Hodel)

offshoots came from a grove being cleaned out near what used to be Jensen's Date Gardens in Indio. By 2000 when they had matured and were bearing fruit, Adair could see that they were not Halawy, although the palms and fruit resembled those of that variety. Adair suspects that they were seedlings from a Halawy grove and were mistaken for Halawy offshoots. Because the fruit had good flavor and showed promise, Adair retained the palms and continued to grow them on. He currently has 12 palms, all with offshoots. He named the variety in honor of one of his customers, Mariana Chaffee. Fruit of Mariana are similar in size and appearance to those of Halawy but have slightly thicker skin, are much tastier in the *khalal* stage, and have a more "buttery" flavor in the *tamar* stage.

An attractive soft date of good quality, Mariana fruit handle and store well and do not "sugar up" much in storage. Adair sells some of his crop of Mariana in the *khalal* stage because, unlike Halawy fruit, fruit of Mariana in this stage are crunchy and sweet. In this regard, *khalal* fruit of Mariana are similar to those of Barhee. Mariana fruit have shown little or no checking and not much drop or souring, but their susceptibility to other damage from rain or humid weather is unknown. Halawy is one of the most problem-free dates and Mariana is similar to it in this regard. Adair primarily sells Mariana at farmers' markets in Southern California.

Fruit. *Khalal* yellow, *rutab* amber (Figure 36), *tamar* light or golden brown with very little glaucous bloom; oblong with rounded apex; 37–45 mm × 21–27 mm; skin thin, shrinking with flesh but with moderate to heavy blistering; flesh soft, more or less translucent, caramel-like, more or less chewy, brownish outer zone, amber inner zone, little rag; flavor pleasing, somewhat "buttery"; early ripening.

Seed. Grayish, mottled with tan to brown; narrowly oblong-elliptic, apex pointed, base more or less rounded and very slightly pointed; 17–25 mm × 4–7 mm; germ pore toward base; furrow medium–wide, shallow.

Date Culture And Management

We have borrowed liberally from Dowson (1982), Nixon and Carpenter (1978), and Zaid (2002) to summarize information on the culture and management of commercial date palms. Abdul-Baki et al. (1998) provided an excellent summary of soil, water, and fertilizer management in date orchards. We have confined our discussion to in-field culture and management. Consult Nixon and Carpenter (1978) and Rygg (1975) for further detailed information on the physiological and chemical development of date fruit, packing house management, and postharvest handling and storage of dates.

CLIMATE

Date palms grow best and produce the highest-quality fruit in regions with a prolonged hot dry summer and without rain or humid weather during the fruit ripening period in late summer and early fall (Abdul-Baki et al. 1998; Aslan et al. 1991; Zaid 2002). The total heat requirement ranges from 4,000 to over 4,500 degree-days, a calculation based on the sum of the number of degrees of daily mean temperature above 64.4°F (Aslan et al. 1991). Rain or humid weather during fruit ripening damages and spoils fruit, leading to serious losses. Date palms can tolerate temperatures as high as 125°F and as low as 20°F for several hours without sustaining serious damage. All of this means that date palms are well adapted to the Coachella Valley in California's low desert.

Date palms can also be grown in areas farther to the southeast, such as the Imperial and Bard Valleys in California and the Phoenix and Yuma areas in Arizona. Because of increased rain and more humid weather during fruit ripening in these areas, though, production is more problematic there and growers must carefully select varieties with fruit that are resistant to damage from rain or humid weather.

Date palms are commonly grown in subtropical and tropical regions around the world, such as Los Angeles and Miami (USA), Sydney (Australia), and Singapore, but in areas of cooler temperatures, increased rain, or high humidity, production of marketable fruit is impossible and the palms serve primarily as ornamental landscape plants.

SOILS

Although a deep, sandy loam is best (Abdul-Baki et al. 1998; Aslan et al. 1991), date palms can be grown on a wide variety of soils so long as they have good water-holding capacity, are well drained, and have a deep profile that allows water to penetrate readily to 6 to 8 feet.

Date palms tolerate alkaline or saline soils better than most plants, but growth and fruit quality are reduced under very saline conditions (Zaid 2002).

PROPAGATION

Date palms can be grown from seed, offshoots, and tissue culture. About half of the date palms grown from seed will be staminate (male) and produce only pollen, while the other half will be pistillate (female) and produce only fruit and seed. Date palms originating from seed are not true to type: every seedling tree is different from every other, and few produce good-quality fruit. Date palms grown from seed that do have outstanding features or produce good-quality fruit can be propagated from offshoots. Because offshoots are genetically identical descendants of the parent palm, they perpetuate any desirable traits. Date palms grown from offshoots of a single palm are actually clones, but they are popularly known as varieties. The primary value of growing date palms from seed is that it is a way to develop new varieties possessing desirable characters or traits.

Seed

Plant fresh seed with the fruit or pulp removed, 1 to 2 inches deep in a well-aerated medium composed of two parts sand and one part organic matter (such as

Figure 37. Date palm offshoots are tied up prior to removal. (Donald R. Hodel)

peat moss or compost) and keep moist but not soggy-wet. Plant in nursery rows or use deep containers (at least 6 inches deep). Use a greenhouse that will allow you to maintain the air temperature between 80° and 85°F or plant the seed when the weather begins to warm up in spring. Seed should germinate in 1 to 2 months.

When the young plants form their first strap-like leaf, transplant them out of community containers and into individual containers filled with a medium similar to that used for germination.

Offshoots

Select offshoots from the parent palm when they have roots, have themselves begun to produce a second generation of offshoots, and have been in contact with irrigated soil for at least a year. Offshoots 6 to 14 inches in diameter (Hodel and Pittenger 2003b; Nixon and Carpenter 1978), weighing at least 25 pounds (Nixon and Carpenter 1978), and having at least 25 roots and 20 leaves (living or dead) (Hodel and Pittenger 2003a) establish and survive best.

On offshoots with poor root development, particularly those higher up on the mother palm, you can mound soil around the base of the offshoot for a year before removal to stimulate root development. Remove and plant offshoots in late spring when soil temperatures have begun to warm up, optimally to 70°F or above. Commercial operations often transplant offshoots later than this, typically in June and July, because that is when there is a break in other date orchard operations.

Leaves should not be pruned from an offshoot until it has been removed from the mother palm. If there are too many offshoots and they are crowded, however, prune out leaves on the smaller offshoots to retard their growth until the larger offshoots reach at optimal size for removal.

The successful removal of offshoots requires a level of care and skill that is only gained through practice. It is a two-person operation. Here are some basic points to consider:

1. Irrigate several days prior to removal to ensure optimal moisture levels in the plant and help soil to adhere to the root ball.

2. Prune the offshoot before removal by cutting and removing old leaf bases and older, lower leaves close to the fiber on the basal 2 to 4 feet of the offshoot. Retain the 10 to 12 newest green leaves, tying them tightly together 2 to 3 feet above the brown fiber (Figure 37). Cut and remove the terminal part of these leaves that extends beyond the tie.

3. Carefully dig soil away from the base of the offshoot, exposing the connection to the mother palm but leaving 2 to 3 inches of soil around the remainder of the base to protect the root initiation zone.

4. Use a specially designed, heavy chisel with a flat, rectangular blade 9 × 4.5 × 1 inches on an iron handle 48 inches long and 1.5 inches in diameter. This chisel should be made of high-quality, tempered tool steel with the blade edges beveled on one side.

5. Clean away old leaf bases and loose sheathing fibers to expose the connection with the mother palm.

6. Place the chisel at the connection at about a 30° angle off the vertical, the flat side of the chisel toward the offshoot and the beveled side toward the mother palm. One worker positions and holds the chisel while the other drives it with an 8- or 10-pound sledge hammer to sever the connection (Figure 38). Remove the chisel by working it up and down parallel to the cutting blade

Figure 38. A two-person team carefully removes offshoots using a heavy sledge hammer and a specially designed, heavy chisel made of high-quality, tempered tool steel. (Donald R. Hodel)

while pulling steadily outward. A single cut may be sufficient but one or more cuts from each side may be necessary. Do not pry the offshoot from the mother palm before the connection is severed. Chisels have recently been developed with an incorporated slide hammer, making the operation safer and somewhat easier than with a sledge hammer, although the new method provides less flexibility in positioning the chisel.

7. After removal, prune off excess old leaf bases (Figure 39).

8. Keep the root ball moist after removal. Plant it immediately, if possible. If you are unable to plant the offshoot immediately, cover its root ball with soil, sphagnum moss, or burlap, and keep the covering moist.

9. Handle the offshoot gently: dropping or jarring may damage the center bud.

10. Plant the offshoot to the depth of its greatest diameter, usually about 14 to 20 inches, packing the surrounding soil to ensure removal of all air pockets.

11. Wrap the newly planted offshoot with burlap or a layer of organic material such as cornstalks to protect against sun the first summer and cold the first winter, but keep the top open to allow for new leaf growth.

12. Irrigate immediately and keep the soil moist but not soggy-wet. Frequency and amount of irrigation depend on weather conditions and the soil type. Monitor soil moisture carefully.

Figure 39. Once removed, offshoots are trimmed again of old leaf bases. (Donald R. Hodel)

13. Offshoots should be well rooted by the end of the first summer. New leaf extension of at least 10 inches indicates a successfully rooted and established offshoot.

Figure 40. Offshoots are frequently planted initially as close as 10 × 10 feet (note how remaining leaves have been tied up and the tips cut off). After two to three years, offshoots will have grown sufficiently so that some can be culled to give the proper permanent spacing in the row. The culls are then used to establish new rows. (Donald R. Hodel)

You can plant offshoots directly into the field in permanent positions, usually at a spacing of 30 by 30 feet, or much closer together in temporary positions, spaced 10 by 10 feet in a nursery row for later culling. In this case, culls are not discarded but transplanted into new, permanent positions in new rows (Figure 40). Losses are nearly always incurred when planting offshoots, making it necessary later for growers to replant to fill gaps in the rows; for this reason, growers might do well to consider planting offshoots at first into 15-gallon nursery containers. Once well rooted, the container-grown offshoots could be planted into the field in permanent positions.

Starting the offshoots in nursery containers is also advantageous because the offshoots can then be grouped together in a nursery setting where growers have more control over light, water, temperature, and humidity conditions, all factors that are critical to the rooting and establishment of offshoots, and in this way growers can reduce their losses. Also, well-rooted, container-grown offshoots nearly always establish rapidly and successfully in the field because they are subjected to little or no root disturbance, thus eliminating the need to fill gaps in rows later on.

Tissue Culture

Tissue culture or *in vitro* micro-culture has been used with some success to propagate date palm varieties, although plants propagated in this way can be difficult to transfer to a soil medium and are subject to heritable genetic mutations. Date palms derived from tissue culture are not utilized to any extent in the United States, though, primarily because a sufficient supply of offshoots is usually available and there is some concern about the genetic instability of tissue-cultured plants. In some other date-growing countries, though, tissue culture has resulted in a shift to varieties like Barhee and Medjool, largely because too few offshoots are available and these two varieties are perceived to be superior to others that are available. Zaid (2002) provides an extensive review of tissue culture propagation of date palms.

WATER AND IRRIGATION

Adequate irrigation is critical for optimal date palm growth and production. Sufficient water must be applied to compensate for evaporative loss from the soil, meet palm requirements, and leach salts contained in the irrigation water beyond the root zone. The frequency and amount of water depend on the soil type, weather conditions, and method of application.

The practical rooting area of mature date palms is 15 feet deep or more and out to 10 feet or more from the trunk, and date palms extract 90 percent of the water they use from the top 5 feet of soil (Zaid 2002). This makes localized irrigation systems such as micro-sprinklers or drip irrigation more efficient than non-localized systems such as flood or furrow irrigation. Despite this, most date palms in the United States are still flood irrigated (Figure 41). Recently, though, some growers have successfully employed micro-sprinklers. As water becomes more scarce and expensive in California and Arizona, drip or micro-sprinklers are becoming a more attractive alternative to flood irrigation. Drip irrigation and micro-sprinklers allow growers to be more precise and accurate

Figure 41. Most commercial date orchards in the United States are flood irrigated. (David Karp)

in how much water they apply and where they apply it, and the results are more-efficient irrigation and less waste. Table 8 summarizes critical characteristics of some types of irrigation systems for date palms.

Any irrigation system for date palms should be designed and operated to keep soil moist to a depth of 7 to 8 feet. Use a soil auger or other suitable instrument to monitor water movement in the soil. About 4 to 6 inches of water at each irrigation may be sufficient once the soil has been moistened to the desired depth. Irrigation systems should not deposit water on the leaves or trunks as it can lead to disease.

Historic applications for fully bearing date palms in the Coachella Valley show that on light sandy soils they require about 9 to 12 acre-feet of water annually and 12 to 18 inches per month in the summer. Growers typically flood-irrigate bearing palms on lighter soils every 7 to 14 days in the summer and every 20 to 30 days in the winter. Growers using micro-sprinklers generally irrigate on shorter intervals.

More-frequent but lighter irrigations may be necessary for small, newly planted small palms and offshoots because they have shallower, less-extensive root systems than mature, bearing palms. Lengthen the time between irrigations but increase the amount of water applied at each irrigation as the palms grow larger and come into bearing age.

Growers can also access on-line historic or real-time evapotranspiration (ET) data from the California Irrigation Management Information System (CIMIS) at http://www.cimis.water.ca.gov/ or the Arizona Meteorological Network (AZMET) at http://ag.arizona.edu/azmet/ for help in scheduling irrigations more accurately based on the formula

Table 8. Characteristics of various irrigation systems for date palms

Characteristic	Type of irrigation system			
	Flood	Furrow and basin	Sprinkler	Micro (sprinkler or drip)
Application affected by wind/temperature	no	no	yes	yes
Easy to apply	yes	yes	yes	yes
Easy to automate	no	no	yes	yes
Efficient use of water	no	somewhat	yes	yes
Installation costs	low	medium	high	high
Interferes with mechanical operations	no	yes	yes	yes
Labor intensive	yes	yes	no	no
Operational costs	low	low	high	low
Suitable for small palms	yes	yes	no	yes
Use on all soil types	no, not good on very sandy soils	yes	yes	yes, but requires greater density of emitters on very sandy soils
Use on sloping terrain	no	yes, but not on steep slopes	yes	yes
Water must be clean	no	no	no	yes

$$ET_{crop} = ET_o \text{ x } Kc$$

where: ET_{crop} = Crop Evapotranspiration (inches per day)

ET_o =   Reference Evapotranspiration (inches per day)

Kc   =   Crop Coefficient

Crop evapotranspiration is the water need of the particular crop—in this case, date palms. Reference evapotranspiration is the water need of a standard reference crop, usually a tall fescue grass maintained in optimal condition at each of the more than 100 computer-driven, computer-connected weather stations throughout California, and available to growers at the CIMIS Web site. Crop coefficient is the water needs of a particular crop, in this case date palms, in relation to that of the standard reference crop. Crop coefficient values for optimal yield of date palms range from 0.8 to 1.0 (Zaid 2002).

Although the date palm is remarkably salt tolerant (Furr and Ream 1967), excessive salt accumulation from irrigation water can reduce growth and fruit production and quality (Abdul-Baki et al. 1998; Aslan et al. 1991; Zaid 2002). If salt accumulation is verified, apply two or three heavy irrigations in rapid succession in the winter to leach salts to beyond the root zone. Water containing high amounts of sodium may eventually produce a soil structure that inhibits water penetration. If so, apply 1 to 3 tons of gypsum (calcium sulfate) per acre, working it into the soil before the heavy irrigations, to help promote leaching (McGeorge 1954). On calcareous soils, apply 400 to 1,200 pounds of sulfur per acre, working it into the soil several months before the heavy leaching irrigations (McGeorge 1955). Winter cover crops can improve water penetration.

Some growers reduce or withhold irrigation during the harvest season to make it easier to get into the orchard, promote the drying of some soft dates, or reduce fruit drop in humid weather. Irrigation can be greatly reduced or omitted for 2 to 3 months or even longer on Deglet Noor or Khadrawy on deep soils of high water-holding capacity without reducing fruit yield or quality if ample moisture has been provided up to early August (Furr and Armstrong 1955; Furr, Currlin, and Armstrong 1952; Furr et al. 1951).

Nevertheless, unnecessarily long intervals between irrigations are not recommended: they can cause problems with salt removal and soil management once irrigation is resumed. Also, it is critical that soil moisture be adequate for palm growth in the spring and early summer, as a deficiency then hastens fruit ripening and reduces fruit size and quality (Aldrich 1942).

FERTILIZER

In most cases date palms benefit from fertilizer applications, although there seems to be a lack of consensus and relatively little published information concerning types of fertilizers as well as rates and frequency of application, especially in California. Zaid (2002) based his fertilizer recommendations on nutrients lost or "exported" through fruit and pruned leaves. Haas and Bliss (1935) reported that one acre of date palms loses 25.5 pounds of nitrogen, 4.4 pounds of phosphorus, and 61.6 pounds of potassium annually. Embleton and Cook (1947) estimated that leaf pruning alone resulted in annual per-acre losses of 22 pounds of nitrogen, 1.6 pounds of phosphorus, and 65.1 pounds of potassium. Furr and Barber (1950) estimated that one acre of Deglet Noor loses 171.6 pounds of nitrogen annually.

Using the above reports and estimates, Djerbi (1995) stated that to produce 110 pounds of dates per palm annually, 39.6 pounds of nitrogen, 11.9 pounds of phosphorus, and 71.3 pounds of potassium must be applied to every acre. Assuming a planting rate of 50 palms per acre, that comes to 0.79 pound of nitrogen, 0.24 pound of phosphorus, and 1.43 pounds of potassium per palm annually. Extrapolating from the above data, Zaid (2002) estimated the average annual nutrient loss per palm to be 0.77 pound of nitrogen, 0.20 pound of phosphorus, and 1.19 pounds of potassium, as compared to actual, average, annual, worldwide fertilizer applications per bearing palm of 1.43 pounds of nitrogen, 1.43 pounds of phosphorus, and 1.91 pounds of potassium.

Zaid (2002) recommended application of 1.16 pounds of nitrogen, 0.30 pound of phosphorus, and 1.19 pounds of potassium per palm annually for those palms 4 years of age and older. For palms less than 4 years of age, he recommended similar annual amounts of phosphorus and potassium, but only 0.58 pound of nitrogen per palm, about half the amount recommended for older palms. In contrast, Furr and Barber (1950) reported that 4 to 6 pounds of nitrogen per palm annually was an adequate amount on most soils in the Coachella Valley.

Because relatively little information is available on date palm nutritional needs and fertilizers, it might be helpful to review the nutritional needs of palms that are used as landscape ornamentals. Research in Florida, an area of sandy, alkaline, nutrient-poor soils, shows that a $N:P_2O_5:K_2O:Mg$ ratio of 2:1:3:1 was most appropriate (Broschat 1999). Broschat and Meerow (2000) recommended applying 5.2 pounds of a sulfur-coated, controlled-release "palm special" fertilizer per palm every 2 to 3 months.

Visible symptoms of deficiencies in nitrogen, potassium, and magnesium (the three elements most commonly deficient in landscape palms) are sometimes helpful in diagnosing nutritional disorders, but it is important to recognize that other factors, such as root injury or disease, too much or too little water, excessive salts, cold temperatures, improper pH, or herbicide toxicity, can cause similar symptoms. A nitrogen deficiency appears as a general yellowing or loss of green color over all the leaves. Potassium and magnesium deficiencies occur first on the older leaves, where potassium deficiency symptoms appear as yellow and orange flecking (especially visible when backlighted) and magnesium deficiency symptoms take the form of a yellow band around the outside of the leaf blade.

Before applying corrective fertilizers, make sure to confirm these visible symptoms through laboratory leaf tissue analysis. Abdul-Baki et al. (1998) analyzed pinnae from Deglet Noor and Medjool leaves bearing date fruit bunches in their axils and found nitrogen levels that ranged from 1.16 to 2.01 percent, potassium levels ranging from 0.42 to 1.38 percent, and magnesium levels ranging from 0.14 to 0.27 percent. Broschat and Meerow (2000) recommended using middle pinnae from mid- to lower-middle canopy leaves while Van Zyl (1983) recommended using pinnae from 1- to 3-year-old leaves for nitrogen and magnesium testing and 6-month-old leaves for potassium testing. Normal ranges vary considerably; for the few palms studied they are nitrogen 1.20 to 3.50 percent, potassium 0.85 to 2.75 percent, and magnesium 0.25 to 1.00 percent (Chase and Broschat 1991).

It is probably best to divide the recommended amount of fertilizer into two or three applications during the growing season. Timing of the applications may be important. Zaid (2002) recommended breaking up the annual fertilizer needs into two applications: one at flower initiation (February) and the other at fruit initiation (July). Growers in Israel apply nitrogen fertilizers monthly from November to April and phosphorus and potassium at 3-month intervals throughout the year. Granular fertilizers can be applied at the recommended rate and frequency throughout the irrigated area, whereas water-soluble fertilizers can be injected into the irrigation water.

Although inorganic sources of nitrogen are often used in California, animal manures, such as steer or poultry, have also been used in California and Arizona at rates of 5 to 15 tons per acre per year, applied in the late fall or winter. If a cover crop is used, though, the manure is applied in the spring after the cover crop has been turned under. In most instances, date palms in arid regions have not responded to applications of potassium and phosphorus (Nixon and Carpenter 1978).

Winter cover crops of sourclover, sweetclover, and other legumes have proven beneficial for young date palms. They provide organic matter that decomposes readily, releasing nutrients and also improving water penetration. Hubam sweetclover in particular has given good results in the Coachella Valley when planted in the fall. On good soil, such a cover crop may provide all the nitrogen required by young palms for several years. After 12 to 15 years, when the leaves have reached their maximum spread, the palms generally cast too much shade to allow any useful growth of cover crops (Nixon and Carpenter 1978).

Disposal of date palm greenwaste, primarily old leaf and fruitstalk prunings, is an increasing concern. Technology now allows growers to chip or grind palm leaves, fruitstalks, and even trunks for use as a compost or soil mulch rather than dispose of them in a landfill. Leaf and fruitstalk prunings of mature Deglet Noor contain more organic matter but less nitrogen than a cover crop of sourclover (Embleton and Cook 1947). Greenwaste should be composted properly and applied at an appropriate depth to avoid fire hazard.

PRUNING

Generally it is best to retain all good green leaves on a date palm because bearing capacity corresponds to the number of green leaves in its crown (Nixon 1940, 1943). However, removal of a few green leaves may be justified in some instances. For instance, a few leaves may have to be removed to allow placement of protective bags or covers over the fruit bunches.

Deglet Noor palms 10 to 20 years old can hold so many leaves that a considerable number are below the fruit bunches, increasing the relative humidity and resulting in higher incidences of checking and blacknose (Nixon 1947). These excess leaves also compete with the fruit for water, especially if irrigation is omitted or reduced during harvest as is sometimes the case with Deglet Noor. In these instances, remove just enough lower leaves from the palm to expose the lower ends of most fruit bunches. In varieties like Halawy and Khadrawy with short fruitstalks, the removal of green leaves below the bunches will reduce fruit production.

If green leaves are removed, avoid reducing fruit quality for the current season or the number of fruit bunches for the next season by also removing a corresponding number of bunches to maintain an adequate leaf-to-bunch ratio. Deglet Noor palms in full production and pruned to 100 to 125 leaves (4 to 5 years of growth) can carry one moderately thinned bunch of fruit for each eight to nine leaves.

Remove leaves in June so fruit bunches will be better ventilated during July and August when checking is most likely to occur. The number of leaves should still be sufficient during this period for the palm to accumulate sufficient carbohydrate reserves (Aldrich and Young 1941).

You can remove dead or partly dead leaves at any time, but it is easier to cut them before the base has become hard and dry. Infrequent, periodic freezes can kill all or parts of leaves, in which case it is best to retain all leaves having any green tissue (Nixon 1938a).

Remove spines during the winter from all of the previous year's growth to facilitate pollination and handling of fruit bunches. Use a sharp pruning knife with a long, curved blade on a foot-long handle.

POLLINATION

Date palms are dioecious. For fruit of good quality to be produced, pollen from staminate (male) flowers on staminate palms must move to pistillate (female) flowers on fruit-bearing (pistillate) palms. Wind and, perhaps, insects pollinate dates that grow in the wild. In commercial production, however, artificial pollination is necessary to ensure adequate fruit set.

Staminate flowers can be harvested and held for several days before being used for pollination. If this is the case, they will usually dry and shatter, releasing their pollen (Figure 42). There are then several methods used to pollinate date palms. Collect the released, dry pollen and dust it on cotton balls about 1 to 2 inches wide. Place two cotton balls between the rachillae of a pistillate inflorescence. Pollen can also be applied using a small, manual insecticide duster similar to the puffers used in the poultry industry (Figure 43). Another method is to cut two or three rachillae of staminate flowers from freshly opened staminate inflorescences and place them securely between the rachillae of the pistillate inflorescences during the first 2 or 3 days after they have opened. Tie the pollinated pistillate inflorescence with a slipknot 2 to 3 inches from its distal end to hold the rachillae of staminate flowers in place yet still allow the inflorescence to expand. Successful pollination will result in a fruit set of 50 to 80 percent, which is sufficient for a full crop.

Pollination in Southern California is primarily done in March and April, though it can vary slightly depending on the season and date variety. Pollination during low temperatures early in the season usually yields poor results (Brown, Perkins, and Vis 1969). Under these conditions, take long paper bags of the type used for French bread and securely fasten them over the flower clusters at pollination to improve fruit set (Reuther 1946).

Some growers have pollinated date palms mechanically using a tractor-pulled, motorized, modified agricultural duster (Perkins and Burkner 1973). This method usually produces a poorer fruit set than hand pollination, but yields and fruit quality end up being about the same because the number of fruit per bunch is similar to that of hand-pollinated, thinned bunches. In many cases, a grower will employ both mechanical and hand pollination methods to ensure an optimal fruit set.

Figure 42. Cut staminate inflorescences hang in a drying shed so the pollen will drop to the floor where it can be collected for hand pollination. (David Karp)

Figure 43. Date palms are hand pollinated by applying a puff of pollen to the pistillate inflorescence. (David Karp)

Collect the pollen by cutting staminate inflorescences early in the morning, as soon as possible after the enclosing bract has burst open. Occasionally, growers will secure bags over unopened inflorescences to preserve pollen and reduce losses if they cannot collect pollen daily. Some experienced growers are able to cut open the staminate inflorescence before it opens on its own and collect the pollen. Usually, when you press the middle or lower part of an unopened bract between your thumb and forefinger and you hear a crackling noise, the flowers inside are mature and pollen is ready for release.

Pollen that you do not use immediately should be allowed to dry to prevent losses due to mold. Avoid heat above 80°F during the drying process. Handle small lots of pollen by cutting off rachillae with mature flowers and spreading them in a thin layer on paper in shallow trays. For large quantities, spread rachillae on wire screening shelves or on trays with wire screening bottoms, with a container positioned beneath to catch the pollen that falls over several days. Mechanical pollen extraction and collection equipment is available to speed up the collection process and actually removes about 40 percent more pollen than manual extraction (Burkner and Perkins 1975).

Pollen will quickly lose viability if subjected to moisture or heat (Gerard 1932). In contrast, dry pollen stored at room temperature will keep satisfactorily for 2 to 3 months. Pollen can be stored satisfactorily for even longer periods, even a year or more, if dried, placed in airtight containers, and held below 40°F. Place the pollen in an open jar within a larger airtight container that has kiln-dried calcium chloride, silica gel, or some other suitable desiccant in the bottom. One pound of calcium chloride is adequate for 5 pounds of pollen (Aldrich and Crawford 1941).

There was a time when the source of pollen was not considered critical and pollen was collected indiscriminately, mostly from seedling staminate palms. It is now common practice, however, to select staminate palms that have desirable pollen characteristics and propagate them vegetatively from offshoots the same way that fruit-bearing or pistillate date palms are propagated. Staminate varieties in the United States include Mosque, Medjool BC3, Deglet Noor BC4, Fard No. 4, Jarvis No. 1, Boyer No. 11 (Zaid 2002), and Crane (Figure 44), although growers do not always recognize them. One staminate palm will provide sufficient pollen for 50 pistillate palms (Zaid 2002).

Several characters are useful when evaluating the suitability of a staminate palm for pollen production:

- **Time of bloom.** The staminate palm must bloom and release its pollen as early as the flowers of pistillate palms become receptive. Cultural techniques can affect bloom time, so grow pollen-bearing palms on the south side of the orchard and under the same favorable conditions as the fruit-bearing palms.

- **Number and size of inflorescences.** Fewer staminate palms are required if they have more and larger inflorescences.

- **Flowers and pollen.** Staminate flowers that adhere to the rachillae after they are harvested without shedding their pollen are best, especially if they are to be used immediately in pollination. Flowers should have abundant pollen and, preferably, petals that do not open completely when the enclosing bract first splits.

Figure 44. Crane is one of the few staminate varieties of date palm cultivated solely for its pollen. (Donald R. Hodel)

- **Compatibility.** Some pistillate varieties set fruit better using pollen from certain staminate palm varieties than pollen from others.

- **Metaxenia.** The direct effect of pollen on the fruit is called *metaxenia*. To some extent, pollen from some staminate palms can affect the size of fruit and seed and the time of ripening more than that from others (Nixon 1928, 1935, 1936). Pollen has less effect on fruit size, though, than thinning has, but in some instances pollen that promotes early ripening can be used to a grower's advantage (Whittlesey 1933). In date-growing districts where late-ripening varieties might be subject to inclement weather or where there is insufficient summer heat for proper maturation, the use of pollen (such as that of the Fard variety) to promote earlier ripening has given good results.

FRUIT THINNING

Fruit thinning is necessary to increase fruit size, improve fruit quality, prevent delayed ripening, reduce the weight and compactness of fruit bunches, and ensure adequate flowering the following year (Nixon and Crawford 1937). Reducing the number of fruit per bunch (per fruitstalk or infructescence), called bunch thinning, or reducing the number of bunches per palm, called bunch removal, are techniques used to thin fruit.

Bunch Thinning

Thin every bunch by reducing either the number of fruit per rachilla or the number of rachillae per bunch, resulting in the removal of not less than one-half and not more than three-fourths of the total number of flowers or fruit.

In varieties with fewer, especially long rachillae, such as Deglet Noor, cut the tips of all rachillae back to remove one-third or slightly more of the total number of flowers at pollination. Also, cut out entire rachillae from the center of the bunch, enough to remove one-third of the total number from most bunches and one-half from very large bunches, about 6 weeks after pollination. Deglet Noor palms thinned in this manner will normally carry at maturity 20 to 35 fruit per rachillae and 30 to 50 rachillae per bunch. Bunches average 15 to 25 pounds of ripe fruit each.

In varieties with shorter but more numerous rachillae, like Halawy and Khadrawy, cut back the tips of the rachillae only enough to even up the end of the bunch at pollination (usually removing about one-tenth to one-sixth of the flowers or fruit) and cut out one-half or slightly more of the rachillae, all from the center of the bunch 6 to 8 weeks after pollination. If growers use spreader rings in the center of bunches, they usually omit cutting back the tips of rachillae and instead do all the thinning by cutting out more rachillae from the center of the bunch, about two-thirds of the total number.

Because it is time consuming and costly, the removal of a certain number of flowers or fruit from individual rachillae instead of cutting back rachillae is seldom practiced in semidry dates such as Deglet Noor. Growers of fancy soft dates, though, especially very large ones such as Medjool, do employ this method (Nixon 1951, 1956b).

Bunch thinning should be done as early as possible to maximize the benefit to fruit size. Cut the rachillae back at pollination and remove rachillae from the center of bunches 6 to 8 weeks after pollination.

The grower must determine the exact method and amount of bunch thinning after considering the variety, relative importance of size in grade, local weather conditions, and potential effects on the fruit from different degrees and methods of thinning. Consider the following points when determining how much to bunch thin and in what way (Nixon 1940, 1942, 1956a; Nixon and Crawford 1942).

- Reducing the number of fruit per bunch increases fruit size and quality.

- Reducing the number of fruit per rachilla is slightly more effective in increasing size than reducing the number of rachillae.

- Cutting back the tips of rachillae increases the fruit's susceptibility to checking, blacknose, and, particularly in some of the softer varieties, shrivel of ripe fruit, much more than cutting out a comparable amount of entire rachillae.

- Overthinning increases puffiness and blistering in the fruit.

- Slightly larger fruit are produced on the outer rachillae of a bunch than on the inner ones.

- The larger the bunch, the more fruit it can satisfactorily carry.

- The earlier thinning is done, the more effective it is in increasing fruit size.

- If damp weather occurs during ripening, fruit rot and souring are more likely in large bunches than in small ones.

- In some instances it may be beneficial to set an upper limit on the size of the bunch, regardless of the amount of thinning required.

- Thin all bunches uniformly in order to obtain uniform size and quality.

- Monitor thinning judiciously: small differences in thinning can result in significant changes in fruit quality. For example, a difference of only 2 inches in the rachilla length of a large Deglet Noor bunch may mean the difference between one-third and one-half of the total number of flowers or fruit. Removing one-half of the flowers may result in 15 to 25 percent more blacknose than removing only one-third.

Bunch Removal

Sometimes it is desirable to thin fruit by removing entire bunches. If a palm is allowed to carry too much fruit one year, it may not produce enough flowers for a normal crop the following season. The amount of fruit that a date palm can safely carry depends on the palm's age, size, vigor, and variety, the number of green leaves it holds, and various cultural practices.

Although the number and size of leaves are the best indicators of palm vigor and fruiting capacity, it is not a simple matter of working out leaf:fruit ratios. The value of a given leaf to the palm declines with its age, and no two leaves are the same age (Aldrich et al.

1942; Nixon 1947). For example, a four-year-old leaf of Deglet Noor is only about 65 percent as efficient in photosynthesis as is a one-year-old leaf (Nixon and Wedding 1956). Generally, though, Deglet Noor palms in full production and pruned to 100 to 125 leaves (4 to 5 years of growth) can carry one moderately thinned bunch of fruit for each eight or nine leaves. Based on this ratio, the safe load for such a palm would be 12 to 15 bunches, or perhaps 2 to 3 more on a palm that is exceptionally vigorous and with larger-than-average leaves.

The total number of leaves on a date palm is easy to estimate because of their orderly arrangement. Leaves are usually arranged in nearly vertical but slightly spiraling columns. Multiply the number of leaves in one column by the number of columns to give an approximate number of leaves on the palm. Estimating the number of leaves can still be tricky, though, because sometimes it is difficult to count the number of columns of leaves on a palm.

Sometimes a grower will remove a bunch that is small, such as the first or last bunch of the season. It may not be economical to pollinate, thin, and bag small bunches with less than 6 or 8 pounds of fruit if the number of large bunches is already sufficient to carry all the fruit that the palm should bear.

Also, it is a good practice to remove bunches from an offshoot for the first three years after planting, since at that stage the growth of leaves and roots for establishment is more critical than fruit production. Leave one or two bunches of fruit the fourth year and then progressively increase that number each year until you reach full production in 10 to 15 years.

BUNCH MANAGEMENT

Commercial growers typically pull each bunch down through the leaves after the pollination season and tie the peduncle to one of the lower leaves (Figure 45). This practice helps prevent fruit scarring and reduces the chance that a bunch will break off as its weight increases. If you are going to cut out the center rachillae from the bunches, the time to do it is during bunch pull- and tie-down. The peduncle continues to elongate after tie-down and the bunch may move, so it is often necessary to adjust the ties about a month later. This is also the time to pull down and tie smaller and later bunches that were not extended far enough at the time of the first tying.

Timing for bunch pull- and tie-down is important. Do not pull down and tie bunches until the peduncle is long enough to distribute the weight of the fruit evenly along its length. If you wait too long to pull and tie down bunches, though, you can increase the danger of breakage. Even partially broken peduncles can cause significant losses because they are often the source of shriveled or low-grade fruit. Because the peduncle elongates in the first few weeks after pollination and is then rather pliable, bending easily at the base, it is safest to begin pull down and tying about a month after pollination.

Most bunches with long peduncles will come down satisfactorily even without tying if you pull them down or gently flex them as far as possible without danger of breakage, once or twice after pollination. Some bunches may be resistant to pull-down, however, or they may have a tendency to break if pulled down normally. Because bunches typically do not require support until the fruit have attained three-fourths of their full size, "tying up" bunches at that later time may be a more satisfactory alternative than "tying down" at an earlier stage.

On small or young palms, support the bunches to keep them off the ground by securely attaching the peduncle to a sturdy wooden stake.

FRUIT GROWTH AND DAMAGE FROM RAIN AND HIGH HUMIDITY

Occasional wet weather has little effect on fruit in the early stage of development, but high humidity just prior to the *khalal* stage can cause minute superficial breaks in the skin, commonly referred to as checks. The abundance and nature of checking varies with different date varieties. In some varieties such as Deglet Noor, checking is primarily near the tip of the fruit and, when severe, is followed by a darkening and shriveling of the fruit, a condition known as *blacknose* (Aldrich et al. 1946; Haas and Bliss 1935; Nixon 1932, 1933).

Blacknose mars the fruit's appearance, lowers its grade, and may increase culls. Especially important on Deglet Noor, blacknose is primarily related to seasonal conditions that result in high humidity when fruit are in the *kimri* stage (late green), which coincides with the attainment of maximum size. Excessive soil moisture and anything that increases the humidity around fruit during this period can aggravate the condition. Interplantings—even cover crops and tall weeds—can increase the incidence of blacknose unless the fruit bunches are well above the interplantings.

Although checking does not occur once fruit have attained the *khalal* stage, contact with water at this stage can produce deeper and longer breaks or cracks in the skin and even the flesh. This cracking, sometimes called *splitting,* is more severe when it occurs in the latter part of the *khalal* stage. In some varieties such as Deglet Noor, irregular curling back of the skin and outer flesh, known as *tearing,* can accompany severe splitting. If tearing is not excessive, the torn skin and flesh may revert to its normal position once dry weather returns, but still the grade of the fruit is lowered. Humid weather during the *khalal* stage also promotes the development of fungi that cause serious spoilage

Figure 45. Commercial growers typically pull bunches down through the leaves after the pollination season and tie the peduncle to one of the lower leaves to prevent fruit scarring and reduce the chance that a bunch will break off as its weight increases. (Donald R. Hodel)

from rot. Provide proper ventilation of bunches and avoid overthinning to reduce the incidence of checking and blacknose.

Once the fruit flesh softens in the *rutab* stage, the skin no longer breaks readily from contact with water or high humidity, but instead the fruit absorb moisture and tend to become sticky, making them less attractive and more difficult to handle. The increase in moisture content and interference with normal drying or curing promotes fermentation and souring, which can result in serious losses.

Rain and high humidity cause little damage once the fruit have attained the *tamar* stage unless they are exposed to prolonged contact with water and absorb excessive moisture.

PROTECTING FRUIT FROM RAIN

In nearly all cases it is best to protect fruit by covering the bunch during ripening, typically once the fruit begin to show *khalal* color. Most commercial growers use light brown kraft paper, wrapping it around the bunch and tying it to the peduncle with the lower end left open (Figure 46). In varieties with a relatively open crown such as Khadrawy and Halawy, a white paper cover causes less sunburn than brown paper (Nixon 1946b, Nixon and Reuther 1947). In the Bard Valley and elsewhere, growers cover Medjool bunches with lightweight cotton bags with the upper portion waterproofed.

Because covers reduce ventilation and so increase atmospheric humidity within the bunch, covering bunches before the *khalal* stage may actually promote checking and blacknose. To lessen the chance of fruit damage under covers, growers can turn under and roll up the sides of covers, allowing better air circulation around the fruit. Then when rain threatens, the covers can be pulled down. Unfortunately, this practice is labor intensive and not practical except on young palms or small plantings.

Bunch ventilation becomes more critical with more frequent rain and periods of high humidity. In such conditions it is best to use a cover that flares out and does not extend down around the sides of the bunch. If you use this type of cover, it may be necessary to protect the bunch with a good grade of porous cloth or netting that will allow good air movement but still exclude birds and insects. Bunch thinning, especially when it involves removal of center rachillae, promotes better air circulation under covers. Bunch thinning is an integral and critical part of the bunch covering strategy.

Spreader rings made of heavy galvanized wire are sometimes used to keep the centers of bunches open. To be effective in promoting ventilation and reducing checking and blacknose, the rings must be inserted before the *khalal* stage. Spreader rings also reduce fermentation and rot in the centers of bunches (Bliss and Bream 1940). Rings are typically about 10 inches in diameter and crimped about every ¾ inch. Inserted in the center of the bunch, the ring provides a structure that the rachillae can drape over to keep them well separated and allow air movement into the center of the bunch. The crimping helps keep the rachillae and ring in place.

Figure 46. Most growers wrap bunches in light brown kraft paper to protect ripening fruit when they begin to show *khalal* color. (Donald R. Hodel)

Figure 47. Growers typically make several pickings to harvest fruit during a season, as shown here on Medjool, because not all of the dates on any one bunch will ripen at the same time. (David Karp)

HARVESTING DATES

Time of Harvest

Harvest season lasts 3 to 4 weeks for early maturing varieties to 2 to 3 months for the late ones. Growers typically harvest fruit in several pickings over a season because not all of the dates on any one bunch will ripen at the same time (Figure 47). Dry dates like Thoory and the semidry Zahidi are usually left until all of the fruit are ripe and then the entire bunch is cut. With Deglet Noor, growers usually cut the entire bunch after all the fruit are ripe and then soften the drier fruit through hydration (Figure 48).

The stage of maturity for handpicking fruit depends on local weather conditions, consumer preference, and date variety. Under favorable weather conditions, leave the fruit on the palm until they reach the stage of maturity at which they are customarily eaten or stored. From a consumer's standpoint, fruit may be considered ripe when it becomes palatable. Fruit may be consumed from the peak of *khalal* stage, when they have attained their most intense red or yellow coloring and maximum

weight, to the final *tamar* stage, when they have lost the most of their moisture and will keep in storage without special attention.

Although in Arabic culture it is customary to eat large quantities of many varieties of dates in the *khalal* stage, most varieties are too astringent at this stage for the American palate. A few varieties, however, such as Barhee, are so palatable in the *khalal* stage that some people, if given the opportunity, might acquire a taste for them. Indeed, a few growers market Barhee and some other varieties in this stage.

Loss of moisture begins with loss of *khalal* color and continues through to the *tamar* stage. Many consumers prefer to eat fruit immediately after the loss of *khalal* color, while it is still plump and the moisture content is very high. Fruit in this condition are fragile and difficult to handle, however, and in order to be marketed they would have to be consumed immediately or placed in storage at a very low temperature. For use as a dessert fruit, most Americans prefer dates after they have passed this plum-ripe stage.

The most desirable stage of maturity for eating dates varies with the variety. Fruit of some varieties have such a high moisture content that there would be little left to

Figure 48. Bunches of fully dried, *tamar*-stage Deglet Noor fruit hang from the tree waiting to be harvested. (David Karp)

Figure 49. Soft fruit are typically placed in shallow trays during and after harvesting, as shown here with Medjool. (David Karp)

Figure 50. Soft fruit of Medjool are spread out in shallow trays to dry in the desert sun. (David Karp)

eat if they were dried sufficiently to keep well without cold storage. Some varieties ferment and sour more readily than others. With the better varieties of soft dates, the best stage for keeping quality comes when the fruit have lost their watery consistency and have become pliable to the touch but not tough. Marketing fruit in just the proper condition requires coordination and cooperation between the grower and packer. In some instances, observation and experimentation are necessary to determine the proper stage of maturity for picking for storage or immediate consumption.

Harvesting Methods

Use shallow trays for collecting soft dates, which are more fragile and require more care in picking and handling than firmer varieties (Figures 49 and 50). Place fruit on the tray no more than two or three layers deep to avoid crushing and bruising. Firmer and semidry varieties may be layered in deeper trays.

Like all other in-tree date palm operations, harvesting becomes more difficult and problematic as palms become older and taller. Saddles, ladders, mobile steel towers, and hydraulically operated lifts with catwalks are used to reach the fruit (Figure 51). Lightweight aluminum extension ladders are favored for palms over 25 or 30 feet high. Many growers permanently attach a straight ladder 10 to 20 feet long immediately below the crown of leaves on extra tall palms to reduce the need to carry extra long ladders around the orchard.

After reaching the crown on a very tall palm, the picker typically uses a belt or saddle. A chain passed over the petioles of three or four green leaves and attached to the belt supports the picker beneath the leaves while he or she braces his or her feet against the palm trunk and picks with both hands. A second or safety chain is placed loosely around petioles of a few lower green leaves to check the picker's fall if the first support fails. The use of long-handled hooks attached to the end of the chains and hooked over the petiole of a green leaf to support the belt facilitates movement around the palm.

Mechanical harvesting systems have been used to harvest semidry date varieties, especially Deglet Noor and Zahidi. These varieties are well adapted to this type of handling and can be harvested in one or two "bunch harvests" (Rygg 1975). Workers are raised into the crown by hydraulically operated towers, lifts, or cherry pickers, and remove entire bunches when most of the fruit on a bunch are mature. After the bunches are brought down from the palms, a mechanical shaker removes the fruit from the bunch and they are run through a separator into pallet bins. The machinery for mechanical harvesting is expensive, though, and requires large acreage to justify its use. Small growers who wish to harvest mechanically will need to form cooperatives or otherwise arrange to share equipment. Bunches can also be vigorously hand shaken to remove fruit (Figure 52).

Figure 51. Many growers harvest dates using hydraulically raised platforms or lifts to reach fruit bunches, as shown here with Medjool. (David Karp)

DISEASES, DISORDERS, AND PESTS

Fortunately, there are no serious diseases or pests of date palms and fruit in the United States. White scale or *Parlatoria* date scale, the most serious and dangerous insect pest of date palms and caused by the scale insect *Parlatoria blanchardii*, was introduced on early importations but was eradicated in 1936 by a cooperative federal and state effort (Boyden 1941). Bayoud, the most serious disease of date palms, caused by a variant of the fungus *Fusarium oxysporum*, has been excluded from the United States by quarantine measures. Another variant of *Fusarium oxysporum*, however, kills the ornamental Canary Island date palm (*Phoenix canariensis*) in coastal areas of California (Hodel 1985), but state quarantine has mostly excluded it from date-growing regions. In their publication, Carpenter and Elmer (1978) provide a comprehensive review of date pests and diseases, while Djerbi (2003) covered diseases and Oihabi (2003) covered insect pests.

Diseases, disorders, and pests cause economic losses in all phases of date production. They reduce yields by destroying or damaging fruit in the field, resulting in direct losses. In the packing house, destroyed or damaged fruit must be culled and downgraded for by-product use, resulting in direct losses due to the additional handling required to clean out the culls. Measures implemented to control diseases, disorders, and pests in the field and the packing house directly increase production costs.

Employ integrated pest management (IPM) principles and practices when managing diseases and pests in the date orchard. Consider using the beneficial insects, cultural practices, and less-toxic pesticides emphasized in the IPM approach. Use pesticides only as a last resort, and always follow rules and regulations and product label recommendations governing their use. Consult the University of California IPM Program's online resources (http://www.ipm.ucdavis.edu/) and appropriate professionals or regulatory officials if necessary.

DISEASES

Fruit Rots

Fruit rots, spotting, and dropping caused by fungi can result in considerable fruit damage when humid or rainy weather occurs during the ripening season. Losses can run as high as 10 to 40 percent of the crop (Zaid 2002). You can reduce the chances of damage by providing good ventilation of bunches and by protecting fruit from the rain as described earlier. The judicious use of fungicides during the late *khalal* and *rutab* stages can help to reduce rots. Growers in areas where humid weather is a regular, routine occurrence during the ripening stages can avoid or reduce losses from rots by harvesting fruit just before they mature fully and allowing them to ripen off the palm in warm maturation rooms.

Black Scorch

A fungal disease caused by *Ceratocystis paradoxa* (alternate state: *Thielaviopsis paradoxa*), black scorch damages young palm leaves, stunting, distorting, and blackening them as though they had been scorched by heat. Newly appearing inflorescences and infructescences can also be damaged. Sometimes the disease attacks the center bud and will kill the palm. Palms that recover from a center-bud attack will show a characteristic bend in the region of the infection. Practice good sanitation and promptly remove and destroy or safely dispose of infected plant parts. The application of an appropriate fungicide to the fresh cuts and surrounding tissue has been shown to help reduce the spread of disease.

Diplodia Disease

Caused by the fungus *Diplodia phoenicum*, *Diplodia* disease affects leaves and offshoots. Leaves develop a dull reddish or yellowish brown streak in the rachis and, when severe, the disease can kill offshoots (Zaid 2002). The disease is seldom found in well-cared-for date orchards. Practice judicious sanitation and remove and destroy or safely dispose of infected plant parts. The application of an appropriate fungicide to the fresh cuts and surrounding tissue has been shown to help reduce the spread of disease. Because wounds such as those made during pruning or cutting facilitate entry of the fungus (Zaid 2002), avoid wounding the palm, and make sure to disinfect all pruning and cutting tools with bleach prior to working on each palm.

Graphiola Leaf Spot

Caused by the fungus *Graphiola phoenicis* and sometimes referred to as false smut, *Graphiola* leaf spot forms numerous, dark brown or black, cylindrical protuberances on the leaves from which yellow spore dust escapes. Severe infections may adversely affect the growth and fruiting of date palms by causing premature death of green leaf tissue, but it has not become an economically important disease in Arizona and California because of low humidity in the area.

Omphalia Root Rot

A minor disease caused by two fungi, *Omphalia pigmentata* and *O. tralucida,* and only appearing to cause serious injury to Deglet Noor, *Omphalia* root rot results in the rotting and abortion of roots, followed by loss of vigor, stunting, and eventual failure of the palm to fruit. Although the fungus is widespread in date orchards in California (Kenknight 1948), symptoms are typically associated only with inadequate irrigation or failure to move sufficient water into tight soils.

Rhizosis or Rapid Decline

A minor but fatal disease of unknown cause, rhizosis results in sudden, premature drop of considerable amounts of one-quarter- to three-quarters-developed fruit of vigorous palms in late spring to midsummer. At that stage, unexpanded bud leaves in the center of

Figure 52. Semidry date varieties like Deglet Noor, as shown here, can be harvested in one or two "bunch harvests." After harvesting, bunches can be vigorously hand shaken to remove fruit. (David Karp)

the crown will be faded and wilted and the oldest and lowest leaves will have a reddish brown discoloration. Leaves die rapidly and progressively from the lower part of the crown upward. Although bud rot is usually an early symptom, the last leaves to die are those that emerged the previous year. Fortunately, the disease is relatively rare and has only been observed in palms planted on light soils in the northwestern end of the Coachella Valley, California.

DISORDERS

Black Scald

A condition commonly associated with checking and blacknose but apparently not due to the same cause, black scald results in blackened and sunken areas at the tips or sides of the fruit as if they had been exposed to extreme heat. Its cause is unknown. There is usually a definite line of demarcation between normal and collapsed tissue, and the latter generally has a somewhat bitter taste.

Crosscuts or Transverse Notches

Crosscuts or transverse notching are characterized by abrupt, smooth breaks in the lower part of the peduncle supporting the bunch that look as though the peduncle had been cut with a sharp knife. Fruit on that part of the bunch in line with the crosscut are typically stunted and of poor quality. Often the peduncle will break and all of its fruit may become shriveled and worthless. Crosscuts and a similar condition in leaf bases known as V-cuts result from an anatomical defect in tissues involving internal, sterile cavities that lead to mechanical failure and breaks during elongation of the peduncle and leaf. Crosscuts and V-cuts are more common in varieties that have crowded leaf bases. Occurrence increases as palms become older (Zaid 2002). Sayer and Khadrawy are especially susceptible, although crosscuts seldom occur on the principal commercial varieties. Avoid losses by using non-susceptible varieties or by reducing the number of fruit bunches in susceptible varieties (Zaid 2002).

PESTS

Only insect and mite pests that attack palms or fruit in the field and cause significant damage are discussed here. See Nixon and Carpenter (1978) for a discussion of date fruit pests found in packinghouses and storage. We have borrowed liberally from Davies and Mauk (1997) for much of this discussion of pests.

Generally, pest damage seems linked more to variety than to geography (Davies and Mauk 1997). One basis for classifying date varieties is whether inverted or cane sugar is predominant in the fruit. Khadrawy, Medjool, Zahidi, and other invert-sugar varieties are resistant to mite damage, whereas Barhee and Thoory are somewhat susceptible and Deglet Noor is highly susceptible. Conversely, nitidulid fruit beetle damage is most serious in invert-sugar varieties like Medjool, although Deglet Noor is also susceptible (Davies and Mauk 1997).

Cultural practices that affect pest damage include pesticide usage, bunch bagging and thinning, irrigation, harvest timing, and orchard sanitation. Sulfur, once a reliable control material for Banks grass mites, has now proven inadequate for this pest. Malathion typically provides adequate control for carob moth and nitidulid fruit beetles in normal years, but during wet years infestation rates in orchards treated with this pesticide can still increase three- or four-fold (Davies and Mauk 1997).

Bunch thinning provides better air movement among fruit, minimizing damage from insects, rotting, and souring, and increasing penetration and coverage of pesticides. Thinning may also reduce mite damage by making it more difficult for mites to establish protective webbing. While bagging provides protection from rain, helping to keep fruit dry and reduce their susceptibility to rotting and souring that can attract pests, the kraft paper wraps (bags) themselves can increase mite damage, possibly because of the shade and insulation the material provides. Although Medjool bunches are typically enclosed in cloth bags tied shut at the bottom to help prevent insect access, fallen fruit that collect in the tied end can attract nitidulid fruit beetles that can lay eggs through the cloth, infesting the entire bagged bunch (Davies and Mauk 1997).

Irrigation can affect pest damage in several ways. Mite damage seems to increase with moisture stress. Flood irrigation hastens the decomposition of fallen dates and other debris, reducing host material for carob moths and nitidulid beetles, and is also somewhat effective in controlling gophers. Flood irrigation may also have a negative impact on pest control, however, since flood-irrigated soils remain wet longer, disrupting management operations and encouraging mosquito and eye gnat problems, both local health concerns. While drip irrigation and micro-sprinklers may not disrupt pest-control operations, they can slow the decomposition of fallen dates and actually keep them moist, making them better hosts for insect pests (Davies and Mauk 1997).

Orchard sanitation plays an important role in pest management. Raking or disking fallen dates and other debris eliminates host sites for insect pests. When you control weeds, you allow sunlight to heat the soil and provide a drier environment that is less favorable to pest development.

Banks Grass Mite (BGM)

Emerging as the most serious and damaging pest of date palms, Banks grass mite (BGM) (*Oligonychus pratensis*) feeds on green, immature fruit, causing serious drying and scarring of surfaces and hardening, cracking, and shriveling of flesh (Stickney, Barnes, and Simmons 1950; Elmer 1965). Late-season feeding causes rasping marks or bronzing on the fruit surface. Although the BGM is difficult to see with the naked eye, its fine, dense, heavy web will cover much of the fruit surface and indicate its presence. The webbing collects dust, making even moderate infestations easy to recognize from the ground (Davies and Mauk 1997). The dust-laden webbing creates a barrier to natural enemies and makes it difficult to attain adequate coverage with miticides, thus resulting in optimal conditions for mite population growth (Farrar 2005). Annual losses for 1997 were estimated at $2.5 million plus control costs (California Date Commission 1998).

A particularly aggressive pest, BGM has the ability to quickly and thoroughly colonize and overtake a crop. Mite populations increase rapidly, with a new generation every 7 days (Davies and Mauk 1997). Until recently, applications of sulfur as a dust or spray around June 1 provided control (Elmer 1964, 1965, 1966; Vincent and Lindgren 1958). Growers typically made three to five applications of 50 to 120 pounds per acre from June 1 to late July (Davies and Mauk 1997). For the past few years, however, sulfur has proven inadequate to control this pest. Usage rates for sulfur have soared while control has become nearly impossible, resulting in increasing costs and declining revenues (Perring and Gispert 1997, unpublished work; Mauk 1997, unpublished work).

There is some evidence that sulfur may be suppressing natural enemies of the BGM. Sulfur-treated dates develop larger mite populations and have fewer, if any, beneficial insects as compared with untreated dates, indicating that beneficial insects may play an important role in managing this pest (Perring and Gispert 1997, unpublished work; Mauk 1997, unpublished work).

Little is known about the biology of the BGM. Additional study is needed to develop IPM strategies for this pest. Perring, Gispert, and Farrar (2000) have, however, demonstrated successful biological control of the BGM using an introduced predator mite, and have documented that heavy rain was successful in washing BGMs and their webbing from fruit bunches and suppressing their populations for the remainder of the season. Some growers spray water on the roads around their orchards in an attempt to reduce dust because dust may favor mite development.

Mauk and Bellows (1998) reported somewhat effective control of BGM using Microthiol, a sulfur compound, when applied at 14-day intervals through July. They achieved season-long control of BGM, however, with a relatively new product, Savey, applied at 6 ounces per 250 gallons of water per acre. Because Savey is an ovicide, it is best applied at time of egg laying. It provides inadequate control or none at all if it is applied once the mites have completely colonized the bunch. Mauk and Bellows (1998) also recommended that growers spot-treat areas with sulfur where mites have increased following an application of Savey.

Carob Moth

First appearing in the Coachella Valley in 1982, the larvae of this moth (*Ectomyelois ceratoniae*) feed on fruit, leaving webbing and depositing frass (Farrar 2005, Davies and Mauk 1997). Fruit are most susceptible to attack in late season (August to November) during ripening (Davies and Mauk 1997). Summer rain and high humidity increase surface fungi on fruit that attract moths. Infestation rates are typically 10 percent, but can be as high as 40 percent (Farrar 2005, Davies and Mauk 1997). Deglet Noor appears to be especially susceptible, but other varieties are attacked as well. Annual losses were estimated at $1 million for 1997 plus control costs (California Date Commission 1998).

Growers have relied on treatments of Malathion 5 Dust every 14 to 21 days beginning in late August and continuing through October to control carob moths, typically applying 50 to 100 pounds per acre per application (Davies and Mauk 1997). More-frequent application may be necessary after heavy rains. In dry years, malathion treatments maintained carob moth and nitidulid fruit beetle (see below) infestation rates between 5 and 10 percent, but during wet years rates can triple (Davies and Mauk 1997). Parasitoids of carob moth have not provided economically significant levels of control. The potential for using pheromone-based strategies to disrupt mating appears promising, but this needs more study (Millar and Shorey 1998).

Cultural strategies to minimize damage from carob moths and nitidulid fruit beetles include bagging fruit and bunch thinning, techniques that are typically already employed in the field (Davies and Mauk 1997). Bagging protects fruit from the rain and later spoilage that attracts these pests, while thinning decreases fruit density, providing better air movement to lessen spoilage and to improve malathion penetration and coverage. Because carob moths and nitidulid fruit beetles also feed on waste fruit, orchard sanitation within the date orchard and in adjacent non-date crop areas is highly desirable (Barnes, Warner, and Laird 1984). Remove and dispose of all fallen date fruit, other fruit, and decaying vegetation during the winter, spring, and early summer, times when there are very few dates in the garden on which to feed (Lindgren, Bliss, and Barnes 1948; Davies and Mauk 1997). Also, reduce irrigation to create a drier, less-hospitable environment and harvest the fruit in a timely manner to decrease their exposure to these pests (Davies and Mauk 1997).

Nitidulid Fruit Beetles

Nitidulid fruit beetles cause direct damage by feeding on fruit and indirect damage by vectoring fruit-damaging fungi and bacteria. The two most common nitidulid fruit beetles are the dried-fruit beetle (*Carpophilus hemipterus*) and the corn sap beetle (*Carpophilus dimidiatus*). These and other less-common fruit beetles attack ripe fruit, preferring soft and sour dates. They deposit eggs on or in the fruit and the larvae feed on the fruit until fully grown, when they pupate in the soil. Fruit are most susceptible in late season (August to November) during ripening (Farrar 2005, Davies and Mauk 1997). As is the case with the carob moth, summer rains and high humidity can increase beetle pressure (Nixon 1964).

Although growers do not consider nitidulid fruit beetles as damaging a pest as Banks grass mite or carob moth, they can be locally troublesome in individual orchards and may cause losses equivalent to those caused by carob moth (Davies and Mauk 1997). They can be the most significant pest on high-value varieties like Medjool (Davies and Mauk 1997), which appears especially sensitive (Farrar 2005). Annual losses are estimated at $500,000 for 1997 plus control costs (California Date Commission 1998).

Chemical and cultural control of nitidulid fruit beetles are the same as for carob moth (see above). These beetles are strongly attracted to pheromones and host odors, such as those of rotting fruit, and mass trapping may prove to be an effective control, although more study is needed (Davies and Mauk 1997).

Indian-Meal Moth

Although seldom seen, the Indian-meal moth (*Plodia interpunctella*) can be a serious pest of dates in the packinghouse (Vincent and Lindgren 1972; Carpenter and Elmer 1978; Nixon and Carpenter 1978). However, it is best controlled in the field, before dates enter the packinghouse, using the same management strategies as are used for carob moth.

Raisin Moth

Primarily a field insect, the raisin moth (*Cadra figulilella*) sometimes occurs on dates in the late fall, especially when there are long intervals between pickings (Vincent and Lindgren 1972; Carpenter and Elmer 1978; Nixon and Carpenter 1978). It is best controlled in the field using the same management strategies as are used for carob moth.

Red Date Scale

Exclusively a pest of palms—particularly of date palms, with other palms (Canary Island date palm and California fan palm) as host plants—the red date scale (*Phoenicococcus marlatti*) usually attracts little attention and for the most part goes undetected because it is found mostly under the fiber of overlapping leaf bases, around the base of peduncles, and frequently on roots underground. It often occurs abundantly on the protected white tissue of leaf bases. About the size of a small pinhead and round, individual scales are deep pink to dark red but partly to entirely covered with a white, waxy, cottony secretion that can give them a grayish color with darker centers.

Severe infestations can cover large areas of green leaf tissue, reducing photosynthetic functions. Fortunately, the red date scale is not well adapted to the extreme summer heat and aridity of desert climates, so it is more or less restricted to protected areas and rarely causes appreciable damage under conditions that are favorable for date production. The red date scale might cause more damage on stressed palms, newly planted offshoots, and palms in milder and more humid areas. Practice judicious sanitation and remove and destroy or safely dispose of infested plant parts.

Appendix A.

Roy Wesley Nixon: Professional Biography and Publications

"As for the date work itself it is the best chance in the country for a young man to make a name for himself and participate in the fruits of twenty-five years of the best supported and most successful work yet done in America along the lines of scientific horticulture and rational establishment of a new industry. We are on the eve of a revolution in all agricultural industries through the application of the newest biological science. . . . It seems to me that a young man could not find a more interesting or a more useful field for his life work."

Letter of July 9, 1923, from Walter T. Swingle to Roy W. Nixon regarding the offer of a position as horticulturist at the USDA Date Garden in Indio, California.

Roy Nixon was a Florida boy who went west and made a name for himself as a date palm specialist. Born in 1895 in Dade City, near Tampa, Nixon had an early interest in plants. He worked as a young man for two years in a nursery before moving to Tucson to study agriculture at the University of Arizona, graduating in 1922 with a B.S. degree (Phi Kappa Phi) in horticulture. At Tucson he studied under J. W. Toumey, director of the Agricultural Experiment Station and one of the early promoters of date palm cultivation in the United States. For a brief period after graduation, Nixon was a field inspector in entomology in Arizona, but soon he moved to California where he worked in a nursery in Ontario. In 1923 he joined the USDA in Indio and began an active program of research and writing on various aspects of date palm cultivation that lasted for nearly half a century.

Over his long professional life, Nixon carried out research on dates in the following areas: metaxemia, date varieties, offshoot importation and propagation, blacknose disease, fruit thinning, leaf pruning, ecology, fruit protection, growth regulators, fruit skin separation, mechanical harvesting, and grower relations. His best-known publication was *Date Culture in the United States*, which was published in 1945 and remained the standard reference for decades. It was revised and republished five times (in 1951, 1959, 1966, 1969, and 1978) before going out of print.

In addition to his work in Indio, Nixon's expertise provided opportunities to investigate date growing in other countries. Over 1928 and 1929 he spent six months in Iraq and Iran studying date culture, where he selected 1,300 offshoots for testing in southwestern Texas.

Nixon was awarded a Guggenheim Fellowship in 1948–1949 and he studied the ecology of date varieties in French North Africa. He spent most of his time in Algeria but also did field work in Tunisia, Morocco, and the small date-growing area of southern Spain. In the fall of 1949 he visited the date groves of southern Baja California, which he described as "North America's oldest date gardens." Nixon was invited to Saudi Arabia in 1953, where he spent six months conducting research on date varieties, and then in 1959 he traveled to Libya and Tunisia for two months to attend an international date conference and to investigate cultural practices in those two countries. After his retirement, Nixon took a three-month assignment in 1965–1966 to assist Sudan in promoting more profitable date production methods and then he made a round-the-world trip back home, stopping in Australia where he visited the small date growing area at Alice Springs.

Listed below are Nixon's 79 technical publications on the date palm, excluding a few general, more popular articles he authored. In writing an obituary of Walter T. Swingle in 1952, Nixon recalled the enthusiasm for date palm research he had acquired from Swingle. Without question, Roy Nixon was the most important American date palm scientist of the Twentieth Century. At the height of his career, he was the world authority on dates. Nixon died in 1976 in Indio, where he had lived since 1923. At his funeral a bunch of dates was placed over the casket: a fitting tribute. Nixon was married to Katherine Bogue in 1924 and had two sons, James and Stewart. A very brief obituary by D. H. Mitchell appeared in *Date Growers' Institute Report* 53:3 in 1976.

TECHNICAL PUBLICATIONS ON DATE PALMS BY ROY NIXON

1926. Experiments with selected pollens. Date Growers' Inst. Rep. 3:11–14.

1927. Date pollination experiments in Salt River Valley. Assoc. Ariz. Prod. 6:12.

1927. Further evidence of the direct effect of pollen on the fruit of the date palm. Date Growers' Inst. Rep. 4:7–9.

1928. Immediate influence of pollen in determining the size and time of ripening of the fruit of the date palm. J. Hered. 19(6):240–254.

1928. Pollination experiments in 1927. Date Growers' Inst. Rep. 5:5–7.

1928. The direct effect of pollen on the fruit of the date palm. J. Agr. Res. 36(2):97–128.

1930. Recent observations of date culture in Iraq. Date Growers' Inst. Rep. 7:4–5.

1931. The commercial utilization of differences in time of ripening of dates due to pollen. Date Growers' Inst. Rep. 8:5–6

1932. Date ripening controlled beneficially by using special kinds of pollen. USDA Agr. Yearbook. 1932. pp. 168–169.

1932. Observations on the occurrence of blacknose. Date Growers' Inst. Rep. 9:3–4.

1933. Notes on rain damage to varieties at the US Experiment Date Garden. Date Growers' Inst. Rep. 10:13–14.

1934. The Dairee date: A promising Mesopotamian variety for testing in the Southwest. USDA Agr. Circ. 300.

1934. Recent pollination experiments. Date Growers Inst. Rep. 11:9–11

1935. Bunch thinning experiments with Deglet Noor dates. Date Growers' Inst. Rep. 12:17–19.

1935. Metaxenia in dates. Proc. Amer. Soc. Hort. Sci. 32:221–226.

1936. Further experiments in fruit thinning of dates. Date Growers' Inst. Rep. 13:6–8.

1936. Metaxenia and interspecific pollinations in *Phoenix*. Proc. Amer. Soc. Hort. Sci. 33:21–26.

1937. Discussion of the freeze of January, 1937. Date Growers' Inst. Rep. 14:19-23.

1937. Fruit thinning experiments with Deglet Noor dates. Proc. Amer. Soc. Hort. Sci. 34:107–115. (with C. L. Crawford).

1938. Leaf pruning and fruit thinning following the freeze of January, 1937. Date Growers' Inst. Rep. 15:25–27.

1938. Discussion of the later effects of the freeze of January, 1937. Date Growers' Inst. Rep. 15:27–29.

1939. Effects of certain growth substances on inflorescences of dates. Bot. Gaz. 100(4):868–871. (with F. E. Gardner).

1939. Date growing in the United States. USDA Leaf. 170. 8 pp. (with D. C. Moore).

1940. Fruit thinning of dates in relation to size and quality. Date Growers' Inst. Rep. 17:27–29.

1942. Fruit shrivel of the Halawy date in relation to amount and method of bunch thinning. Proc. Amer. Soc. Hort. Sci. 41:85–92.

1942. Rain and high humidity tolerance of commercial date varieties. Date Growers' Inst. Rep. 19:12–13.

1942. Quality of Deglet Noor date fruits as influenced by bunch thinning. Proc. Amer. Soc. Hort. Sci. 41:103–110. (with C. L. Crawford).

1942. Some factors affecting rate of date leaf elongation. Proc. Amer. Soc. Hort. Sci. 41:77–84. (with W. W. Aldrich [senior author], C. L. Crawford, and W. Reuther).

1943. Flower and fruit production of the date palm in relation to the retention of older leaves. Date Growers' Inst. Rep. 20:7–8.

1944. Dates in the United States. Amer. Fruit Grower. 64(12):9, 24, 26–27.

1945. Date culture in the United States. USDA Circ. 728. 44 pp.

1945. The need for a monograph of the date varieties of the world. Chron. Bot. 9(2/3):153–154.

1946. Bunch protection of the Khadrawy date in relation to sunburn and fruit shrivel. Date Growers' Inst. Rep. 23:10–12.

1946. He brought African dates to Coachella. Desert Mag. 9(4):15–19.

1947. Importations of date offshoots and the men who made them. Date Growers' Inst. Rep. 24:20.

1947. Can a date palm carry too many leaves? Date Growers' Inst. Rep. 24:23–27.

1947. The effect of environmental conditions prior to ripening on maturity and quality of date fruit. Proc. Amer. Soc. Hort. Sci. 49:81–91. (with W. Reuther).

1950. Date culture in French North Africa and Spain. Date Growers' Inst. Rep. 27:15–21

1950. Imported varieties of dates in the United States. USDA Circ. 834. 144 pp.

1950. Observations on the bayoud disease of date palm in Morocco. Plant Dis. Rptr. 34(3):71–73.

1951. Date culture in the United States. USDA Circ. 728. 57 pp.

1951. Date palm: "Tree of life" in the subtropical deserts. Econ. Bot. 5(3):274–301.

1951. Fruit thinning experiments with the Medjool and Barhee varieties of dates. Date Growers' Inst. Rep. 28:14–17.

1951. Leaf characters of the Deglet Noor date palm in relation to age and environment. Proc. Amer. Soc. Hort. Sci. 57:179–185.

1952. Ecological studies of date varieties in French North Africa. Ecology 33:215–225.

1952. Walter Tennyson Swingle (1871-1952). Date Growers' Inst. Rep. 29:21–22.

1953. North America's oldest date gardens. Pac. Disc. 6(1):18–24.

1953. Off-type palms in commercial date gardens. Date Growers' Inst. Rep. 30:8–9.

1954. Date culture in Saudi Arabia. Date Growers' Inst. Rep. 31:15–20.

1955. Size and checking of Deglet Noor dates as affected by fruit thinning and pollen. Date Growers' Inst. Rep. 32:8–10.

1956. Effect of metaxenia and fruit thinning on size and checking of Deglet Noor dates. Proc. Amer. Soc. Hort. Sci. 67:258–264.

1956. How many fruits per strand should be left in thinning the Medjool date. Date Growers' Inst. Rep. 33:14.

1956. Age of date leaves in relation to efficiency of photosynthesis. Proc. Amer. Soc. Hort. Sci. 67:265–269. (with R. T. Wedding).

1957. Difference among varieties of the date palm in tolerance of graphiola leaf spot. Plant Dis. Rptr. 41(12):1026–1028.

1957. Experimental planting of imported varieties of dates in the San Joaquin Valley of California. Date Growers' Inst. Rep. 34:15–16.

1957. Effect of age and number of leaves on fruit production of the date palm. Date Growers' Inst. Rep. 34:21–24.

1958. New root development on trunks of date palms buried in soil. Date Growers' Inst. Rep. 35:6–8.

1958. Morphological effects of specific pollens and thinning on fruit of Deglet Noor dates—a progress report. Date Growers' Inst. Rep. 35:17–18. (with C. A. Schroeder [senior author]).

1959. Effects of gibberellin on fruitstalks and fruits of date palm. Date Growers' Inst. Rep. 36:5–7.

1959. Growing dates in the United States. USDA Infor. Bull. 207. 50 pp.

1960. Gibberellin promising on dates. West. Fruit Grower 13(2):50.

1960. Report of the First International Technical Meeting on Date Production and Processing. Date Growers' Inst. Rep. 37:5–7.

1960. Observations on date culture in Libya and Tunisia. Date Growers' Inst. Rep. 37:22–24.

1960. An unusual disorder of Barhee date palms. Date Growers' Inst. Rep. 37:10–11. (with E. F. Darley [senior author] and W. D. Wilbur).

1961. Skin separation in soft dates. Date Growers' Inst. Rep. 38:10–13.

1961. Quality of Deglet Noor dates in relation to frequency of picking. Date Growers' Inst. Rep. 38:14–15. (with J. R. Furr).

1962. Further studies of frequency of picking in relation to quality of Deglet Noor dates. Date Growers' Inst. Rep. 39:16–18. (with J. R. Furr).

1962. Concentrations of nutrients in pinnae of date palms in relation to an unexplained die-back of leaves in Coachella Valley, California. Date Growers' Inst. Rep. 39:14–15. (with C. K. Labanauskas [senior author]).

1963. Quality of Deglet Noor dates in relation to method of harvesting. Date Growers' Inst. Rep. 40:10.

1964. Viability of date seeds in relation to age. Date Growers' Inst. Rep. 41:3–4.

1964. Rain damage to dates in 1963. Date Growers' Inst. Rep. 41:6–7.

1964. Second report on the bending of tops of Barhee date palms. Date Growers' Inst. Rep. 41:15. (with E. F. Darley [senior author], W. D. Wilbur, and J. B. Carpenter).

1965. Problems and progress in date breeding. Date Growers' Inst. Rep. 42:2–5. (with J. R. Furr).

1966. Growing dates in the United States. USDA Infor. Bull. 207. 50 pp.

1967. Date culture in Sudan. Date Growers' Inst. Rep. 44:9–14.

1969. Growing dates in the United States. USDA Infor. Bull. 207. 56 pp.

1971. Early history of the date industry in the United States. Date Growers' Inst. Rep. 48:26–30.

1978. Growing dates in the United States. USDA Infor. Bull. 207. 63 pp. (with J. B. Carpenter).

1982. Pollination, breeding and selection, Chapter V, pp. 58–78, *in* V. H. W. Dowson, Date production and protection. Plant Production and Protection Paper 35. Rome: United Nations FAO.

Appendix B.

Date Fruit Grading in the United States

Grading standards for Deglet Noor, Halawy, Khadrawy, and Zahidi varieties are presented in Table B-1. Grading standards for Medjool are presented in Table B-2.

Table B-1. Grading standards for Deglet Noor, Halawy, Khadrawy, and Zahidi* dates

Grade	Description
Fancy or A	Whole or pitted dates of one variety; good color; uniform size; practically free of defects; good character; graded score[†] of at least 90.
Choice or B	Whole or pitted dates of one variety; reasonably good color; reasonably uniform in size; reasonably free of defects; reasonably good character; graded score of at least 80.
Choice (dry) or B (dry)	Whole dry dates of one variety; same as above.
Standard or C	Whole or pitted dates of one variety; fairly good color; fairly uniform in size; fairly free of defects; fairly good character; graded score of at least 70.
Standard (dry) or C (dry)	Whole dry dates of one variety; same as above.
Substandard[‡]	Fails to meet requirements of Standard or Standard (dry).

*These varieties are regulated under Date Marketing Order No. 987, as amended.

†*Graded* score (0–100) is the sum of the points given for each of four factors (maximum for each in parentheses): Color (20), uniformity of size (10), absence of defects (30), character (40). Certain limiting rules apply to achieve grades of Fancy, Choice, or Standard.

‡*Below* Substandard are "cull" dates, not included in the classification. Cull dates are those "that are so poorly developed or ripened that they are worthless for human consumption or are dates that have been damaged, contaminated or affected by foreign material . . . which would render then unfit for human consumption. Dates in this group are used in animal feed, the manufacture of alcohol and inedible syrup, and any other outlets not intended for human consumption."

Sources: United States Standards for Grades of Dates, Effective August 26, 1955. Agricultural Marketing Service, USDA, Washington, DC; Dates and Date Products Grading Manual, Food Safety and Quality Service, USDA, Washington, DC, June, 1977.

Table B-2. Grading standards for Medjool dates

Grade	Dates per pound	Description
Jumbo	16–19	No blemishes, skin separation, or dryness
Large	20–23	No blemishes, skin separation, or dryness
Extra Fancy	20–24 (varies)	Minor blemishes; packed all sizes together
Fancy	20-26 (varies)	Some dryness and skin separation; packed all sizes together

Source: Bard Valley Medjool Date Growers Grading Standards.

Appendix C.

Date Palm Germplasm Resources in California and Arizona

Date palm germplasm is being preserved within four formal collections: two in California and two in Arizona. Tables C-1, C-2, C-3, and C-4 list the date varieties, both imported and American, in each collection. Hybrid varieties resulting from breeding experiments are not included in these lists. Table C-5 lists only commercial varieties, showing which of the four collections contain germplasm for those varieties.

Table C-1. Date palm germplasm at USDA–ARS National Germplasm Repository, Thermal, California

Imported and American varieties	Fruit characters	Number of specimens	Provenance of variety
Abada	Soft	1	California
Amir Hajj	Soft	3	Iraq
Ashrasi	Semidry	1	Iraq
Badrayah	Dry	1	Iraq
Barhee	Soft	5	Iraq
Barhee A-19*	n/a	2	California
Bentamoda	Soft	1	Sudan
Boyer #11*	n/a	1	California
Crane*	n/a	1	California
Dayri	Semidry	4	Iraq
Deglet Beida	Dry	1	Algeria
Deglet Noor	Semidry	4	Algeria
Empress	Semidry	2	California
Fard #4†	Soft	1	Oman
Halawy	Soft	4	Iraq
Hayany	Soft	1	Egypt
Haziz	Soft	1	California
Hilali	Soft	2	Iraq
Honey	Soft	2	California
Horra	Dry	1	Algeria
Jarvis #1*	n/a	1	California
Khadrawy	Soft	3	Iraq
Khalasa	Semidry	3	Saudi Arabia
Khir	Soft	1	Saudi Arabia
Khisab	Soft	2	Iraq
Medjool	Soft	3	Morocco
Saidy	Soft	2	Egypt, Libya
Samany	Soft	2	Egypt
Sayer	Soft	1	Iraq
Sphinx	Soft	2	Arizona
T-R	Soft	2	California
Tabarzal	Soft	1	California
Tazizoot BC#3*	n/a	2	Algeria
Thoory	Dry	3	Algeria
Zahidi	Semidry	4	Iraq

*Staminate palm selected as having superior characteristics of early or large pollen production.
†Staminate palm selected for cross with female Fard palm, the latter now unfortunately lost.
Source: Records of Thermal Germplasm Repository, courtesy of Robert R. Krueger, Curator.

Table C-2. Date palm germplasm at USDA–ARS National Germplasm Repository, Brawley, California

Imported and American varieties	Fruit characters	Number of specimens	Provenance of variety
Abada	Soft	2	California
Amir Hajj	Soft	2	Iraq
Ashrasi	Semidry	2	Iraq
Badrayah	Dry	1	Iraq
Barhee	Soft	4	Iraq
Bentamoda	Soft	1	Sudan
Boyer #11*	n/a	2	California
Dayri	Semidry	6	Iraq
Deglet Beida	Dry	1	Algeria
Deglet Noor	Semidry	6	Algeria
Fard #4	Soft	3	Oman
Halawy	Soft	6	Iraq
Haziz	Soft	2	California
Hilali	Soft	4	Iraq
Jarvis #1*	n/a	2	California
Khadrawy	Soft	5	Iraq
Khalasa	Semidry	3	Saudi Arabia
Khir	Soft	3	Saudi Arabia
Medjool	Soft	4	Morocco
Sayer	Soft	2	Iraq
Tazizoot BC#3*	Soft	1	Algeria
Thoory	Dry	4	Algeria
Zahidi	Semidry	4	Iraq

*Staminate palm selected as having superior characteristics of early or large pollen production.
Source: Records of Brawley Germplasm Repository, courtesy of Robert R. Krueger, Curator.

Table C-3. Date palm germplasm at Arizona State University, Tempe, Arizona

Imported and American varieties	Fruit characters	Number of specimens	Provenance of variety
Abada, Black Abada	Soft	3	California
Ashrasi	Semidry	1	Iraq
Badrayah	Dry	1	Iraq
Barhee	Soft	10	Iraq
Bentamoda	Soft	1	Sudan
Braim	Soft	1	Iraq
Dayri	Semidry	5	Iraq
Deglet Beida	Dry	1	Algeria
Deglet Noor	Semidry	2	Algeria, Tunisia
Halawy	Soft	5	Iraq
Hayany	Soft	2	Egypt
Haziz	Soft	2	California
Hilali	Soft	4	Saudi Arabia
Honey	Soft	6	California
Horra	Dry	3	Algeria
Iteema	Soft	2	Algeria
Khadrawy	Soft	3	Iraq
Khalasa	Semidry	2	Saudi Arabia, Oman
Khir	Soft	2	Saudi Arabia
Khisab	Soft	6	Iraq
Maktoom	Soft	1	Iraq
Medjool	Soft	28	Morocco
Menakher	Semidry	1	Tunisia
Peggy Ann	Soft	1	California
Rhars	Soft	1	Algeria
Saidy	Soft	1	Egypt, Libya
Sayer	Soft	2	Iraq
Sphinx	Soft	10	Arizona
Tabarzal	Soft	2	California
Tadala	Soft	1	Algeria
Thoory	Dry	4	Algeria
Zahidi	Semidry	2	Iraq

Source: The Arboretum at Arizona State University (1995) and records of the Arizona State University Horticultural Research Center.

Table C-4. Date palm germplasm at the Yuma Mesa Agricultural Center, Somerton, Arizona

Imported and American varieties	Fruit characters	Number of specimens	Provenance of variety
Barhee	Soft	8	Iraq
Deglet Noor	Semidry	3	Algeria
Hayany	Soft	1	Egypt
Khadrawy	Soft	4	Iraq
Medjool	Soft	31	Morocco
Sayer	Soft	3	Iraq
Sphinx	Soft	1	Arizona
Thoory	Dry	2	Algeria

Source: Yuma Agricultural Center, Somerton, Arizona, Block 13 (November 2001).

Table C-5. Commercial imported and American date varieties cross-listed to germplasm collections

Variety	Thermal, CA	Brawley, CA	Tempe, AZ	Yuma, AZ
Imported varieties				
Amir Hajj	X	X		
Barhee	X	X	X	X
Dayri	X	X	X	
Deglet Noor	X	X	X	X
Halawy	X	X	X	
Hayany	X		X	X
Iteema			X	
Khadrawy	X	X	X	X
Khisab	X		X	
Maktoom			X	
Medjool	X	X	X	X
Samany	X			
Sayer	X	X	X	X
Thoory	X	X	X	X
Zagloul				
Zahidi	X	X	X	
American varieties				
Abada	X	X	X	
Black Beauty* (China Ranch)				
Blonde Beauty				
Brunette Beauty				
8-Ball*				
Empress	X			
Honey	X		X	
Mariana*				
McGill's No. 1				
Sphinx	X		X	X
T-R	X			
Tabarzal	X		X	

*Undescribed variety.

Appendix D.

Dichotomous Key to the Major and Commercial American and Imported Varieties of Dates

Fruit and vegetative characters are nearly always necessary to distinguish date varieties and should be representative and taken from normal palms. The characters showing the greatest contrast are used in this dichotomous key. Keys of this nature have limitations because all characters tend to be somewhat variable. An asterisk (*) after a variety name signifies that the identity of that variety is questionable. In most instances of questionable identity, records or labels have been misplaced or mixed up, making correct identification difficult if not impossible even though the palm appears to be that named variety. A variety name enclosed in double quotation marks ("") signifies that the name is known to be incorrect. In most cases, the labeled palm bears no resemblance whatsoever to the true variety of that name and is obviously mislabeled. A letter "A" denotes American varieties and a letter "I" denotes imported varieties.

We have adapted this key from the key in Nixon (1950), which included imported varieties that the author considered to be "commercial and minor." We have added American varieties that he considered to be "major" in his unpublished manuscript (Nixon 1955). We have also added three imported varieties (Khisab, Samany, and Zagloul) that Nixon (1950) considered to be "other varieties" and one American variety (Abada) that he considered to be a "minor variety" in his unpublished manuscript. We made these additions because we consider these four to be commercial varieties. We were unable to add two American varieties (McGill's No. 1 and T-R) that we consider to be commercial varieties because there was insufficient information about them in Nixon's unpublished manuscript.

1a. *Khalal* fruit yellow.
    2a. Fruit usually not more than twice as long as wide.
        3a. Flesh of ripe fruit very dry, often hard.
            4a. Fruit nearly round ....Jauzi (I)
            4b. Fruit +/- oblong.
                5a. Calyx 3-cleft.
                    6a. Ripe fruit light brown or light grayish brown...Thoory (I)
                    6b. Ripe fruit dark purplish drab. .... Horra (I)
                5b. Calyx margin rounded or only slightly broken.
                    7a. Ripe fruit light buff or pale straw.. Mesh Degla* (I)
                    7b. Ripe fruit dull grayish buff or reddish brown. .... Kenta (I)
        3b. Flesh of ripe fruit not very dry and hard.
            8a. Neck of spines less than 3 cm long.
                56a. Fruit very large (mostly 45-60 x 25-35 mm)
                    57a. Fruit ripen midseason, rag considerable; seed oblong; neck of spines 2-3 cm long. ....Samany (I)
                    57b. Fruit ripen early, rag slight; seed narrowly oblong; neck of spines lacking. ....Tabarzal (A)
                56b. Fruit small to large (mostly 23-52 x 17-30 mm).
                  9a. Spines usually less than 25 in number.
                    10a. Scurf on petiole edges moderate to heavy.
                      11a. Pinnae drooping moderate to pronounced; whitish cast to foliage. ....Maktoom (I)
                      11b. Pinnae drooping slight or none; no whitish cast to foliage. ....Dubayni (I)
                    10b. Scurf on petiole edges sparse or absent.
                      12a. Terminal pinna longer than pinnae just proximally.
                        13a. Pinnae long; drooping pronounced. .... Areshty (I)
                        13b. Pinnae short to medium long; drooping slight or none. ....Khadrawy (I)
                      12b. Terminal pinna not longer than pinnae just proximally.
                        14a. Foliage with pronounced whitish cast..Kush Zebda* (I)
                        14b. Foliage without whitish cast.

15a. Fruit ripen very late..Hilali (I)

15b. Fruit ripen very early to midseason.

16a. Fruiting rachis very short.. Khir* (I)

16b. Fruiting rachis medium long to long.

17a. Fruit nearly black when ripe, ripen very early................ Ammary (I)

17b. Fruit amber to reddish brown when ripe, ripen midseason.

18a. Leaf base fibers in solid bands. .... Amir Hajj (I)

18b. Leaf base fibers not in solid bands. ....Kustawy (I)

9b.   Spines usually more than 25 in number.

19a. Scurf on leaf base edges heavy.

51a.   Spines 40-50 in number, a-r divergence 30-50°. ...."Boo Fagoos" (I)

51b.   Spines 36 in number, a-r divergence 20-25°...Honey (A)

19b. Scurf on leaf base edges absent, slight, or rarely only moderate.

20a.   Terminal pinna usually at least half as long as longest lateral pinna.

21a. *Khalal* fruit orange-yellow...Saidy (I)

21b. *Khalal* fruit light yellow.

50a.   Pinnae BSI 55-70%. .... Haziz (A)

50b.   Pinnae BSI 30-50%.

22a. Fruit broadly ovate or somewhat rounded. ....Barhee (I)

22b. Fruits +/- oblong.

55a.   Neck of spines less than 2 cm long; pinnae stiff. .... Halawy (I)

55b.   Neck of spines 2-3 cm long; pinnae drooping slightly... Sultany (A)

20b.   Terminal pinna less than half as long as longest lateral pinna.

23a. Apical divergence of pinnae less than 60°. ....Iteema (I)

23b. Apical divergence of pinnae more than 60°.

52a. Fruit thickest at middle..Honey (A)

52b. Fruit thickest above or below middle.

24a. Fruit thicker above middle.

25a.   Ripe fruit firm. .... Zahidi (I)

25b.   Ripe fruit soft.

45a. Spines 46 in number....................................Amber Queen (A)

45b. Spines 28-36 in number. ....Bent Keballa (I)

24b. Fruit thicker below middle.

26a.   Fruit slightly constricted near apex; pinnae drooping slight to

moderate. ....Baydh Hammam (I)

26b.   Fruit not constricted near apex; pinnae drooping

pronounced. ....Tazizoot (I)

8b.   Neck of spines more than 3 cm long.

27a. Rachilla tip free of flowers or fruit for 3 cm or more..Khalasa (I)

27b. Rachilla tip not free of flowers or fruit for 3 cm or more.

28a. Fruit oblong-obovate... Apdandon (I)

28b. Fruit oblong, tapering to sharp-pointed apex. .... "Seewah" (I)

2b. Fruit usually more than twice as long as wide.

29a. Neck of spines more than 3 cm long.. Rhars (I)

29b. Neck of spines less than 3 cm long.

30a. Spines less than 30 in number.."16-23" (I)

30b. Spines more than 30 in number.

46a. Pinnae groups distinct except a few above midblade.

49a. Spines 49 in number, neck 1 cm long. ....Blonde Beauty (A)

49b. Spines 34-38 in number, neck lacking. .... Empress (A)

46b. Pinnae groups indistinct above midblade.

31a. Spines very long (>15 cm)... Tadala (I)

31b. Spines short to medium long (<15 cm)..Tafazween (I)

1b. *Khalal* fruit +/- red.

32a. *Khalal* fruit entirely or dominantly red or pink.

    33a. *Khalal* fruit light red or pink.

        53a. Fruit light red; spines 40-60 in number, occupying ¼-⅓ of blade length. ....Deglet Noor (I)

        53b. Fruit pink; spines 38 in number, occupying ⅕ of blade length...Lindy (A)

    33b. *Khalal* fruit medium, deep, or dark bright red.

        34a. Fruit less than twice as long as wide.

            35a. Scurf on leaf base edges medium heavy to heavy.. Kush Batash* (I)

            35b. Scurf on leaf base edges slight, inconspicuous, or absent.

                58a. Fruit rounded oval, ripen very late. .... Khisab (I)

                58b. Fruit ovate, oval, oblong-oval or oblong-obovate; ripen early, midseason, or late.

                    48a. Spine rachis angles not more than 15°. ....El Toby (A)

                    48b. Spine rachis angles not less than 20°.

                        54a. Fruit widest near middle...Sphinx (A)

                        54b. Fruit widest above or below middle.

                            36a. Fruit widest above middle...Tozer Zaid Khala* (I)

                            36b. Fruit not widest above middle..Hamraya-2* (I)

        34b. Fruit usually about at least twice as long as wide.

             37a. Flesh of ripe fruit very soft.

                47a. Fruit deep red, ripen early or midseason.

                    59a. Fruit ripen midseason. .... Zagloul (I)

                    59b. Fruit ripen early.

                        60a. Fruit with little rag; seed narrowly oblong-elliptical...Abada (A)

                        60b. Fruit with considerable rag; seed oblong...Hayany (I)

                47b. Fruit medium red, ripen late...Brunette Beauty (A)

             37b. Flesh of ripe fruit firm.

                38a. Spines less than 30 in number. ....Dayri (I)

                38b. Spines more than 30 in number.

                    39a. Calyx somewhat flattened. ....Menakher (I)

                    39b. Calyx prominent...Hamraya-1* (I)

32b. *Khalal* fruit with some red but yellow background more prominent.

    40a. Perianth set in slight depression.

        41a. Scurf on leaf base edges heavy, conspicuous. ....Ashrasi (I)

        41b. Scurf on leaf base edges light, inconspicuous. .... Medjool (I)

    40 b. Perianth not set in slight depression.

        42a. *Khalal* fruit only very slightly astringent. ....Braim (I)

        42b. *Khalal* fruit considerably astringent.

             43a. Spines less than 25 in number. .... Sayer (I)

             43b. Spines usually more than 25 in number.

                44a. Scurf on leaf base edges moderate.."Banquet Maktoom" (I)

                44b. Scurf on leaf base edges absent. .... "Beach's 8-4" (I)

APPENDIX E.

MAJOR, MINOR, AND OTHER NON-COMMERCIAL IMPORTED VARIETIES OF DATES

The non-commercial, imported varieties of dates are presented in this table. They are grouped into three categories, major, minor, and other, and are arranged alphabetically within each group. An asterisk (*) after a variety name signifies that the identity of that variety is questionable. In most instances records and/or labels have been misplaced or mixed up, making correct identification difficult if not impossible even though the palm appears to be that named variety. A variety name enclosed in quotation marks ("") signifies that the name is known to be incorrect. In most cases the labeled palm bears no resemblance whatsoever to the true variety of that name and is obviously mislabeled. The recognized or possible meaning, if known, is provided below the variety name. The person responsible for introducing the variety (if known) is given in parentheses after the date of introduction. Fruit color is given sequentially in *khalal* stage (non-ripe), *rutab* stage (ripe), and *tamar* stage (cured). Yields, where known, are per palm per year. The state(s), Arizona, California, and Texas, where the variety was introduced or primarily grown and its presence in the germplasm collections at Brawley and Thermal, California, and Tempe, Arizona, are provided in the "Notes" column. All information is taken and adapted from Nixon (1950) and updated where appropriate. Many of these varieties might no longer exist but the descriptive information may allow identification of an unrecognized variety.

Table E-1. Major varieties of dates

Varietal name and meaning	Synonyms	Origin, date of introduction (and introducer)	Distinguishing characters	Fruit	Notes
Khalasa *quintessence* (Popenoe 1913b) *choice* (Dowson 1939)	Khalaseh, Khalasi, Khalas, Khulas, Khlas	Oman Saudi Arabia 1913 (Popenoe)	Slightly to moderately arched leaves with short, stiff pinnae at tips but long, slightly drooping pinnae at base; neck of spines 5 cm long or more; distal 3 cm of rachillae bare of flowers; fruit with unusually small perianth.	Soft; 30–40 × 19–23 mm; oblong-oval; yellow, ripening deep amber, curing light amber to reddish brown with light bloom; flesh tender, melting, translucent, little or no rag; flavor rich, delicate; ripens midseason; yield 125–150 pounds.	Highly esteemed for its fruit, reputed to be the most delicious in the world; has received much lavish praise for excellent quality and flavor; fruit cure and keep well although sometimes consumed in the *khalal* stage; fruit appearance suffers from rain but fruit are not subject to damage from humid weather. Arizona, California. Cultivated at Oasis Date Garden in Coachella Valley, California, but fruit are not sold, rather given to friends and people in the trade (Hansen 2004). In Brawley, Thermal, and Tempe germplasm collections.
Kustawy *having a stone* (Dowson 1939)	Kustawi, Khastawi, Khustawi	Iraq 1902 (Fairchild)	Vigorous growth; full spreading crown of slightly to moderately arching leaves with yellowish cast.	Soft; 28–38 × 19–23 mm; oblong-ovate or oblong-oval; yellow, ripening light brown, curing reddish brown; flesh caramel-like, very little rag; flavor honeylike, pleasing; ripens midseason; yield 150–200 pounds.	Was principal soft date in northern Iraq; fruit are similar to those of Khadrawy but are smaller and skin has strong tendency to blister or separate from flesh, although planting on heavier soil, thinning fruit heavily, and allowing fruit to cure under uniformly controlled humidity tend to mitigate these drawbacks; fruit are little damaged from rain or humid weather. Arizona, California.
Rhars *robust* (Anon. 1920)	Ghars	Algeria 1900 (Swingle)	Leaves rather strongly recurved in distal third; pinnae broad, stiff; fruit narrowly oblong-obovate, very early ripening.	Soft; 45–55 × 20–24 mm; narrowly oblong-obovate; yellow, ripening amber, curing reddish brown; flesh melting, little or no rag; flavor rich, sweet, rather cloying; ripens very early; yield 200–250 pounds.	Was principal soft date in Algeria; susceptible to serious checking, splitting, and souring from rain or humid weather; frost sensitive. Arizona, California. In Tempe germplasm collection.
Saidy *Said* is native name for Upper Egypt (Popenoe 1913b)	Saidi, Sayd, Sayeh, Siwah, Seewah, Sewi, Siwi, Wahi, Oga de Bedrichen	Egypt 1901 (USDA)	Unusually heavy trunk; stiffly spreading leaves rather strongly arched in distal half; pinnae broad, stiff; old, green leaves tend to bend at base and hang down near trunk; *khalal* fruit orange-yellow.	Soft; 35–48 × 22–28 mm; broadly oblong-oval to oblong-ovate; orange-yellow, ripening light brown, curing dull reddish brown; flesh somewhat firm, white or buff colored, very little rag; flavor heavily sweet, rich; ripens late; yield 200–250 pounds.	Was principal variety in oases of Libyan desert; European explorers much praised it for excellence of fruit; flavor improves with storage; fruit are large and attractive in appearance, pack and keep well, but are subject to souring and drop from calyx end rot. Arizona, California. In Thermal and Tempe germplasm collections.
Tazizoot	Tazizaout, Tazizaut, Tazizaoot, Tazezait, Tazerzait	Algeria 1904 (USDA)	Rather striking because of deep green leaves with little flexibility except at the tip, and long, narrow, drooping pinnae that lend a somewhat filigree look to center of crown.	Soft; 40–50 × 20–25 mm; oblong-elliptical; yellow, ripening amber, curing reddish brown with rather heavy bloom lending a bluish cast; flesh melting, usually with much rag; flavor rather "heavy" and syrupy; ripens early; yield 200–250 pounds.	One of the minor varieties of Algeria, seldom very common; fruit are large, attractive, but are usually rather coarse in texture, lack delicacy of flavor, and are more subject to souring, dropping, rot, and checking from rain or humid weather; fruitstalks are too short and erect for easy protection of bunches against rain; palm produces relatively few offshoots. Arizona, California.

Table E-2. Minor varieties of dates

Varietal name and meaning	Synonyms	Origin, date of introduction (and introducer)	Distinguishing characters	Fruit	Notes
Ammary meaning uncertain, perhaps *the abundant* (Popenoe 1913b) *cultivated prosperous*	Amari, Ammaree	Tunisia 1905 (Kearney)	Broad, heavy fruitstalks; fruit ripen very early.	Soft; 34–38 × 18–20 mm; oblong-oval to oblong-obovate; yellow, ripening and curing to nearly black; flesh with some rag, slightly mealy consistency; flavor mediocre; ripens very early.	Common in Tunisia and Algeria; fruit are earliest ripening of any imported variety, usually maturing in early July; perhaps good for home use in areas with insufficient heat to mature other varieties. Arizona, Texas.
Apdandon* *the luster of the tooth*	Ap-e-dan-dan*, Hurshud*, Hurshut*	Pakistan (Baluchistan) 1902 (Fairchild)	Neck of spines very long, 5–13 cm; large pinnae in proximal portion of blade.	Soft; 36–42 × 19–23 mm; oblong to oblong-obovate; yellow, ripening amber, curing reddish brown; flesh tender, melting, with very little rag; flavor delicate, pleasing; ripens early.	Reputed to be one of the best soft dates of Pakistan; fruit are subject to only moderate checking from rain; probably the same as Hurshut and possibly the same as Began Jangi, neither of which is cultivated in the U.S. Arizona.
Areshty *the feathery* (Popenoe 1913b)	Areshtee, Arechti, Arishti, Archeti, Arashti, Arichti, Rishti	Tunisia 1905 (Kearney)	"Feathery" appearance of crown; pronounced curvature of leaves near tip; long, narrow, drooping pinnae often split along midvein.	Semidry; 38–52 × 23–27 mm; oblong-obovate; yellow, ripening amber, curing light reddish brown with light bloom; flesh firm, considerable rag becoming harder with time but never as hard as those of Deglet Noor; flavor mildly sweet; ripens midseason.	Found at oases in Tunisia; also reported from Algeria; fruit are subject to considerable checking, splitting, rot, and souring from rain or humid weather. Arizona, California.
Ashrasi meaning uncertain, perhaps *hard* (Dowson 1939); *tall growing* (Popenoe 1913b)	Ascherasi, Asharasi	Iraq 1902 (Fairchild)	Palm similar to Kustawy but differs in its more vigorous growth; heavy scurf on edges of leaf bases extends along rachis and on spines and abaxial midvein of pinnae; *khalal* fruit yellow with reddish brown stippling around coral-red-tinged perianth; fruit ovate to somewhat wedge shaped.	Semidry; 35–42 × 25–32 mm; ovate to somewhat wedge shaped; yellow with more or less fine reddish brown stippling around perianth, ripening dull amber with some light brown areas, softer portions curing reddish brown; flesh firm, smooth, becoming very hard, considerable rag; flavor rich, nutty; ripens midseason.	Excellent semidry date but seldom sets fruit satisfactorily and tends to shed immature fruit, but planting on heavy soil and prompt pollinating immediately after bracts open tend to mitigate these drawbacks; fruit are resistant to damage from rain or humid weather. Arizona, California. In Brawley, Thermal and Tempe germplasm collections.
"Banquet Maktoom"	——	Iraq? 1913? (Popenoe?)	Palm very similar to Maktoom but differs in its dark maroon leaf bases; slightly less scurf on edges of leaf bases and proximal portion of rachis, especially on young leaves; more spines; larger percentage of double-veined apical pinnae; less-glaucous leaves, more open crown; lack of strips or bands of fibers across leaf bases; and longer, narrower fruit, *khalal* fruit red over yellow background.	Soft; 36–42 × 20–24 mm; oblong-oval; yellow with some faint reddish cast near base, ripening amber, curing dull reddish brown; flesh with moderate amount of tender rag; flavor sweet but cloying; ripens late.	Was thought to have been an accidental import of an unknown variety; confused with Maktoom because palm is similar, but fruit are dissimilar and of inferior quality. Arizona, California.
Baydh Hammam *pigeon egg* (Kearney 1906)	Bayd Hamam	Tunisia 1905 (Kearney)	Peculiar and distinctive oval to obovate fruit with abrupt narrowing or constriction near the somewhat pointed apex as though it had been pinched.	Soft; 34–40 × 20–24 mm; oval to obovate with conspicuous constriction near apex; yellow, ripening amber, curing reddish brown; flesh slightly grainy with thin layer of rag; ripens midseason.	Although said to be highly esteemed in Tunisia, quality of fruit produced in United States is inferior; fruit are subject to moderate damage, mostly checking, from rain or humid weather. Arizona.
"Beach's 8-4"	——	Iraq 1913 (Popenoe)	Palm similar to Dayri but differs in slightly more yellowish green, less-glaucous leaves; leaf curvature less accentuated near tips but more prominent in distal third of blade; pinnae more evenly and symmetrically placed in distal half of blade; and fruit softer, smaller, less red in *khalal* stage but darker when cured.	Soft; 36–44 × 19–23 mm; oblong-oval; yellow overcast with red, ripening dull brown, curing dark brown to nearly black; flesh with rather tough rag; flavor sweet but quickly cloying, flat, lacking character; ripens midseason.	Has been confused with Khalasa and Sayer, but fruit are inferior to both and neither palm nor fruit has anything in common with Khalasa; fruit are darker and less uniform than those of Sayer; fruit set easily and uniformly. Arizona, California.
Bent Keballa meaning uncertain, perhaps *daughter of the south, kissable maiden, the shriveled* (Popenoe 1913b); *daughter of the midwife*	Bent Kebela, Bint Khabala, Bint Qabbaleh, Bint Qabilah, Bint Qibleh, Bint Qubaleh	Algeria 1904 (USDA)	Similar to Iteema but differs in more lush growth; denser crown; pinnae with larger rachis angles, more pronounced drooping, and more widely open folds; fewer and more slender spines; glabrous fruitstalks; and more obovate fruit with milder, less cloying flavor.	Soft; oblong-obovate; yellow, ripening amber, curing reddish brown; flesh melting, very little rag; flavor rich, pleasing; ripens midseason.	Although very attractive and of good quality as a fresh date, fruit are difficult to handle, do not cure well, and are very susceptible to splitting and souring during humid weather. California.

Varietal name and meaning	Synonyms	Origin, date of introduction (and introducer)	Distinguishing characters	Fruit	Notes
"Boo Fagoos" *father of the cucumber* (Kearney 1906)	Algerian Iteema	Algeria (Johnson)	Palm similar to Iteema and Rhars but differs in rather heavy scurf on the edges of leaf bases and proximal portions of rachis and very early ripening fruit; fruit of Rhars ripen only a little ahead of "Boo Fagoos" but are larger, narrower.	Soft; 38–50 × 21–25 mm; oblong, widest a little above middle; yellow, ripening amber, curing light reddish brown; flesh caramel-like, often slightly grainy, thin layer of sometimes tough rag; flavor good but lacking in character; ripens very early.	Large, meaty, soft date of attractive appearance but lacks rich flavor of similar Iteema; fruit subject to checking, and after rain or humid weather considerable rot and souring can occur; early ripening in Coachella Valley results in puffy, hard-to-handle fruit; was grown under name "Algerian Iteema" in Arizona; neither palm nor fruit correspond very closely to true Boo Fagoos of Algeria and Tunisia, which is not cultivated in the U.S. Arizona, California.
Braim	Buraym, Brim, Brem, Brehm, "Berhi"	Iraq 1902 (Fairchild)	Bright green leaf bases and proximal portion of rachis combined with long spine area and moderate leaf curvature; yellow *khalal* fruit with fine red stippling have very little astringency or tannin flavor.	Soft; 34–40 × 20–23 mm; oblong-oval, sometimes slight ovate or obovate; yellow with some fine red stippling, ripening dull brown, curing deep reddish brown; flesh melting, only a trace of rag; flavor delicate, pleasing; ripens early.	Erroneously referred to as "Berhi" (a synonym of Barhee) in Arizona; good soft date, but fruit are small if not heavily thinned, can shrink considerably in curing, and are somewhat susceptible to damage from rain or humid weather. Arizona, California. In Tempe germplasm collection.
Dubayni *name of an oasis, Dubai, near Baghdad* (Popenoe 1913b)	Dubaini, Deboeni, Deboweni, "Burni"	Iraq 1913 (Popenoe)	Rather open center of crown due to widely spaced groups of narrow proximal pinnae; heavy scurf on edges of leaf bases.	Soft; 36–42 × 22–26 mm; oblong-oval or oblong-ovate; yellow with minute reddish brown stippling, ripening dull amber, curing reddish brown; flesh coarsened by considerable rag; ripens midseason.	Was common but not abundant around Baghdad, Iraq; grown under name "Burni" in Arizona; fruit resemble those of Kustawy but are larger, more puffy, coarser in texture, and more susceptible to considerable losses from souring and dropping during rain. Arizona, California.
Hamraya-1* *red* (Kearney 1906)	——	Algeria 1900 (Swingle)	Long leaves with slight to moderate curvature increasing near tip; pinnae crowded toward tip of blade; crown rather open in center due to long spine area; *khalal* fruit dull red.	Firm; 42–50 × 18–22 mm; oblong-elliptical; dull red, ripening dull amber, curing deep reddish brown with moderate bloom lending purplish cast; flesh with white zone around seed; flavor mediocre, often with faint trace of tannin or slightly disagreeable aftertaste; ripens midseason.	Second-class date; fruit are subject to checking and considerable losses, mainly from souring, during rain; one or more varieties named "Hamraya" occur in all principal date-growing regions of the Old World, all have red *khalal* fruit; similar to Fteemy from Tunisia, which is not cultivated in the U.S. Arizona, California.
Hamraya-2* *red* (Kearney 1906)	——	Algeria 1904 (USDA)	Long, only slightly arched leaves; more or less evenly set pinnae, these stiff but subject to some bending and thus appearing shorter; fruit are similar to those of Kush Batash but ripen earlier and are deeper red in *khalal* stage; lack of conspicuous scurf on edges of leaf bases and proximal portion of rachis and abrupt transition from spines to pinnae also distinguish it from Kush Batash.	Soft; 30–38 × 18–24 mm; oval to ovate; deep red, ripening and curing maroon to nearly black with moderate bloom lending purplish cast; flesh melting, if not fully ripe a little coarse from rag; flavor sweet but lacking character; ripens very early.	Fruit somewhat lacking in quality compared to those of Khadrawy, which they resemble in size and shape; very early ripening fruit are subject to much damage from summer rain in Arizona; one or more varieties named "Hamraya" occur in all principal date-growing regions of the Old World, all have red *khalal* fruit. Arizona, California.
Hilali *of the new moon* (Dowson 1939); *moonbeams* (Popenoe 1913b)	Hellali	Oman (Musqat) 1902 (Fairchild)	Stiff, rather short leaves; stiff, broad, short to medium-long pinnae becoming quite short at tip of blade with wide apical divergence; fruit very late ripening.	Soft; 30–36 × 25–27 mm; broadly oval or somewhat obovate; yellow, ripening amber, curing deeper brown; flesh smooth; flavor mild, delicate; ripens late.	Rare and reputed to be highly esteemed in Oman; fruit of good quality, outstanding in their late ripening, trees frequently carrying fruit into January and February; fruit are subject to moderate damage from rain. Arizona, California. In Brawley, Thermal, and Tempe germplasm collections.
Horra *pure or noble* (Kearney 1906); *the free one*	Harra, Herra, Hourra, Hurra, Hurrah	Tunisia 1905 (Kearney)	Stiff leaves and pinnae; the latter set at rather wide angles on rachis lend a bristly appearance; dry fruit are dull purplish.	Dry; 37–45 × 19–23 mm; oblong ovate, tapering from somewhat flattened base to rounded apex; yellow, ripening and curing dull purplish; flesh becoming more or less hard, dry, with thick white central zone; flavor marred by tannin; ripens midseason.	Widely grown in Algeria and Tunisia where it is reputed to be the largest and finest of any dry dates, but in the U.S. fruit retain astringency and quality is diminished; fruit color and appearance are less attractive and fruit are more subject to splitting in *khalal* stage than those of Thoory; fruit are susceptible to date mites but not subject to much damage from humid weather. Arizona, California. In Thermal and Tempe germplasm collections.

Varietal name and meaning	Synonyms	Origin, date of introduction (and introducer)	Distinguishing characters	Fruit	Notes
Jauzi *walnutlike* (Dowson 1939)	Jauzi Ahmar, Juzi, Jozi, Jozee	Iraq 1913 (Popenoe)	Broadly ovate to nearly round, dull amber fruit with hard, dry, firm flesh; fruit of variety Takermest are round but larger than those of Jauzi and ripen nearly black.	Dry; 28–35 × 22–26 mm; broadly ovate or nearly round; yellow, ripening and curing dull amber sometimes with straw-colored area at base; flesh firm, becoming hard and dry, with thick, white central zone; flavor good, rather nutty if fully ripe, otherwise somewhat astringent; ripens midseason.	One of the oldest varieties in Iraq; mediocre, second-rate dry date distinctive in its walnutlike shape; another variety of the same name but from southern Iraq has soft fruit and is not cultivated in the U.S.; fruit are resistant to damage from rain, but humid weather can cause dark spots. California.
Kenta* meaning uncertain, perhaps *the vigorous* (Popenoe 1913b)	Kanta*, Kentah*	Tunisia, Algeria 1905 (Kearney)	Delicate, moderately arched leaves with flexibility increasing distally; narrow, fairly long, slightly drooping pinnae; fruit similar to those of Thoory but smaller and also similar to those of Mesh Degla and M'Kentichi Degla, but darker, softer, and more finely wrinkled.	Dry; 30–38 × 18–20 mm; oblong to slightly oblong-obovate; yellow with more or less fine brown stippling, ripening and curing dull amber or reddish brown apically, grayish buff basally; flesh becoming firm but seldom hard and brittle; flavor very agreeable, comparable to that of Thoory; ripens midseason.	More common in Tunisia than in Algeria; considered one of best dry dates and has reputation for high yields; fruit are comparable to those of Thoory but smaller, less attractive; fruit are subject to checking but resistant to other damage from rain or humid weather. Arizona, California.
Khir*	——	Saudi Arabia 1904 (USDA)	Short spine area; few spines, arranged singly, gradual transition to pinnae; rachillae abruptly branching from end of peduncle; oblong-spatulate seed; very early flowering.	Soft; 35–42 × 18–21 mm; oblong; yellow, ripening amber, curing chestnut brown; flesh caramel-like, slight and tender rag; flavor mildly sweet; ripens early.	Date of some merit; fruit are resistant to damage from rain or humid weather; very early flowering (often before pollen is available and sometimes when late winter temperatures are sufficiently cold to damage flowers) is a drawback. Arizona, California. In Brawley, Thermal, and Tempe germplasm collections.
Kush Batash* meaning uncertain, perhaps *the sweetmeat* (Popenoe 1913b); *vigorous palm*	Qush Batash	Oman (Musqat) 1913 (Popenoe)	Perhaps could be confused with Hamraya-2, but differs in lighter red *khalal* fruit and slightly later ripening; moderate to heavy scurf on edges of leaf bases and fruitstalks; shorter spine area; and larger spines transition to pinnae more gradually.	Soft; 30–38 × 20–23 mm; oblong-oval or slightly obovate; red with some yellow, ripening and curing deep brownish black with heavy bloom lending light purplish cast; flesh smooth with practically no rag; flavor mild, agreeable, becoming richer and a little cloying after curing; ripens early.	Date of some merit; fruit cure and keep well and are not subject to serious damage from occasional rain. Arizona, California.
Kush Zebda* *the butter date* (Popenoe 1913b)	Qush Zabad	Oman (Musqat) 1913 (Popenoe)	Resembles Maktoom but differs in lack of conspicuous scurf on edges of leaf base and midrib of proximal pinnae; more pronounced whitish cast to leaves; and only slight maroon mottling at base of oldest leaves.	Soft; 28–35 × 19–22 mm; broadly oval; yellow, ripening amber, curing chestnut; flesh melting, practically no rag when ripe; flavor rich, pleasing, with buttery suggestion; ripens midseason.	Reputed to be one of the best soft dates of Oman, but small size is a drawback; fruit are subject to severe checking and splitting, although losses from souring and rot during humid weather have been small. California.
Menakher *the nostrils* (Popenoe 1913b)	Manakhir, Monakhir	Tunisia 1905 (Kearney)	Heavy trunk; striking, deep green leaves; large, oblong, dull deep reddish brown fruit.	Semidry; 45–52 × 22–24 mm; oblong with slightly oblique base; red with some yellow, ripening amber, curing dull deep reddish brown; flesh firm with some seldom-objectionable rag; flavor pleasing, raisinlike; ripens late.	Rare and highly esteemed in Tunisia; large fruit are attractive in appearance and flavor; fruit susceptibility to checking and splitting is about the same as that of Deglet Noor, but fruit are not quite so subject to rot during humid weather. Arizona, California. In Tempe germplasm collection.
Mesh Degla	——	Algeria 1900 (Swingle)	Leaves with slight to moderate curvature, increasing only slightly distally; fruitstalks with moderate scurf.	Dry; 32–40 × 17–20 mm; oblong; yellow, ripening and curing light brown or pale straw-colored; flesh hard, white; flavor insipidly sweet; ripens midseason.	Inferior quality date but popular in Algeria because it produces well under adverse conditions; fruit are very resistant to damage from rain. Arizona.

Varietal name and meaning	Synonyms	Origin, date of introduction (and introducer)	Distinguishing characters	Fruit	Notes
"Seewah"	——	1890	Heavy trunk; deep green leaf base with maroon near fiber; neck of spine 4–6 cm long; oblong, pointed fruit.	Soft; 34–47 × 18–21 mm; oblong, tapering from middle to rather sharply pointed apex; yellow, ripening amber, curing reddish brown; flesh with little rag; flavor rather intensely sweet and cloying; ripens midseason.	"Seewah" is a synonym of Saidy, but neither the palm nor the fruit resemble that variety; a second-rate date; fruit are not especially susceptible to damage from rain. Arizona.
"16-23"	——	?	Medium trunk; medium-long leaves with slight curvature; large, oblong fruit.	Soft; 40–48 × 20–22 mm; oblong; yellow, ripening amber, curing reddish brown; flavor delicate, pleasing; ripens early.	Date of sufficiently good quality to warrant further study; fruit are subject to only slight checking from rain. Arizona.
Tadala	Tadalla, Tedalla, Teddala	Algeria 1900 (Swingle)	Numerous heavy, rather long, closely grouped spines; large, early ripening fruit.	Soft; 46–52 × 18–21 mm; narrowly oblong, tapering from about middle to more or less pointed apex, frequently slightly constricted just above base; yellow with faint orange tinge, ripening amber, curing only slightly darker; flesh with considerable rag; flavor pleasing but not distinctive; ripens early.	Fruit have been compared to those of Rhars but are not equal to it in quality, although they are subject to less damage from rain or humid weather; rather coarse, thin flesh and large seed are drawbacks. Arizona. In Tempe germplasm collection.
Tafazween	Tafazwin, Tafazaouin, Tefzaouine	Algeria 1904 (USDA)	Stiff leaves; numerous, short spines with rachis angles and a-r divergence generally above 45°; long, narrow fruit curing to bright, somewhat translucent, light reddish brown; checking mostly transverse and short, from middle to apex.	Soft, 40–50 × 18–22 mm; narrowly oblong, widest a little above middle; yellow, ripening light amber, curing bright, somewhat translucent, light reddish brown; flesh caramel-like, some tough rag; flavor rather delicate, agreeable; ripens early.	Considered in Algeria to be one of best for planting in saline soils. Arizona, California.
Tozer Zaid Khala*	Tozeur Zid Kahala	Tunisia 1905 (Kearney)	Very similar to Hamraya-2, but differs in larger, narrower pinnae; oblong-obovate fruit with red *khalal* color commonly less uniform; and fruit less subject to damage from rain with checking in longitudinal, apical lines.	Soft; 34–40 × 22–26 mm; oblong-obovate or less commonly oblong-oval; red, sometimes over orange-yellow background toward apex, ripening and curing deep brownish black with light bloom; flesh slightly coarse with very little rag; flavor good; ripens early.	This variety and Hamraya-2 may be only slightly different strains within the same variety; a second-class date subject to checking, but other damage from rain or humid weather is slight. Arizona.

Table E-3. Other varieties of dates

Varietal name and meaning	Synonyms	Origin, date of introduction (and introducer)	Distinguishing characters	Fruit	Notes
Adebet et Teen* *the sweetness of the fig*	——	Algeria 1904	Long, narrow, falcate fruit sometimes with one surface concave; heavy trunk; leaves strongly curved; stiff pinnae and numerous spines set at angles widely divergent with rachis.	Soft; large, long, narrow, falcate or with one surface concave; yellow, ripening amber, curing reddish brown; ripens midseason.	Fair quality date, but tendency to sour easily is drawback. California.
Ahmar Msab* *the red stream*	——	Algeria 1904	Numerous, brown, more or less round spots, 2–4 mm diameter on orange-yellow fruitstalks.	Soft; medium; oblong; dull red, ripening amber, curing reddish brown; ripens early.	Mediocre date; fruit are subject to moderate spoilage from humid weather. Arizona, California.
"Allona"	——	Iraq 1929 (Nixon)	——	Soft; small; oblong, base somewhat flattened; dark red, ripening dull dark brown, curing nearly black; ripens midseason.	Mediocre date; true Allona (not cultivated in the U.S.) is small, amber, soft date similar to Amir Hajj. Texas.
Amhat	——	Egypt 1914 (Mason)	Moderately long leaves only slightly arching; pinnae stiff; bold strips of fiber across the rather broad, yellowish green leaf bases; fairly heavy spines are numerous, very closely set within groups; fruitstalks are orange-yellow.	Soft; small; oblong, broadly rounded apex; yellow, ripening amber, curing reddish brown with moderate bloom lending a purplish cast; flavor sweet, pleasing; ripens midseason.	Reputed to be one of six leading commercial varieties in Egypt and one of the sweetest and most popular varieties for fresh eating; however, it appears poorly adapted to date growing areas in the U.S.; rather small fruit do not cure readily and are subject to damage from rain or humid weather. California.
Amri	Amiri, Amary	Egypt 1901	Palm resembles Hayany but differs in its shorter, slightly more pendulous leaves, stiffer pinnae, and shorter spine area.	Semidry; large; oblong, tapering somewhat to a bluntly pointed apex; red, ripening dull brown, curing a deeper, somewhat purplish brown; flesh coarse, fibrous, with considerable tough rag; ripens midseason.	Rated as an important export variety of Egypt but in the U.S. fruit are coarse, lacking in quality, and subject to considerable damage from checking, blacknose, and rot. Arizona.
Aooshet* *the recurving* (Popenoe 1913b)	Ausheh, Aujeh	Algeria 1904	Broad leaves, stiff proximally, moderately curved in distal one-third; green leaf bases; numerous spines 10-20 cm long with a rather heavy neck 2-3 cm long; stiff pinnae, apical divergence 50-65°; and long, slender to medium-heavy, yellowish green fruitstalks with heavy scurf.	Soft; medium to large; oblong-oval; yellow, ripening amber, curing deep reddish brown; flesh rather coarse, grainy; ripens early.	Fair quality date; fruit are very susceptible to damage from rain, mainly severe souring and checking in long, transverse lines covering the fruit nearly to base. Arizona, California.
"Ashag"	"Russel's No. 28"	Iraq 1913	Unusual appearance with medium-long to long, stiffly spreading, vertically closely spaced leaves; broad, green leaf bases with heavy scurf; very short spine area with few to moderate number of solitary, medium-long spines; and rather stiff, closely and evenly spaced pinnae.	Soft; large; ovate-oblong or tapering from broad base to somewhat pointed apex; yellow, ripening amber, curing reddish brown; flavor mild, pleasing; ripens midseason.	Fruit are similar to those of true Ashag (not cultivated in the U.S.) but differ in lacking rose tinge. California.
Azerza-1*	——	Algeria 1900	——	Soft; large; oblong; deep yellow with sparse brownish red stippling, ripening amber, curing deep reddish brown, nearly black.	Very inferior date; fruit have tough skin, usually blistering and checking badly; fruit seldom cure satisfactorily and are subject to considerable spoilage during humid weather. Arizona.
Azerza-2*	——	Algeria 1900	——	Semidry; medium; oblong; yellow, ripening dull amber, curing dark reddish brown.	Very inferior date; fruit have tough skin, usually blistering and checking badly; fruit seldom cure satisfactorily and are subject to considerable spoilage during humid weather. Arizona.
"Azmashi"	——	Algeria 1913 (Popenoe)	——	Dry; small to medium; oblong, tapering somewhat to a bluntly pointed apex; yellow, ripening and curing dull amber or pale brown; ripens midseason.	Inferior date; true Azmashi is rare, much esteemed, soft date from Algeria and not cultivated in the U.S. California.

Varietal name and meaning	Synonyms	Origin, date of introduction (and introducer)	Distinguishing characters	Fruit	Notes
Badrayah *of Badrah*, an oasis in Iraq near the Iranian border (Dowson 1939)	Bedraya, Bedraihe, Badrahi, Badraihi, Badurahi	Iraq 1929 (Nixon)	Palm has some resemblance to Zahidi but differs in less-glaucous leaves, more yellowish leaf bases with wide bands of fiber across them; more slender spines tending to be in groups of 2; and narrower, less crowded, more drooping pinnae.	Dry; large; oblong with rounded base and apex; yellow, ripening and curing amber at apex, pale faded yellow near base, light bloom; flesh firm, considerable rag, somewhat mealy; flavor sweet, delicate; ripens midseason.	Well known and highly esteemed in Iraq but not widely grown there. Texas. In Brawley, Thermal, and Tempe germplasm collections.
Bahrab	——	Iraq 1929 (Nixon)	——	Soft; small; narrowly oblong, slightly wider above middle; red, ripening amber, curing deep reddish brown, nearly black, bloom at base lending purplish cast; flesh with little rag; ripens very early.	Fair quality date of Fard type, well know near Baghdad, Iraq, but nowhere abundant. California, Texas.
Banawish*	——	Iraq 1929 (Nixon)	——	Dry; medium; oblong; red over dull yellow background, ripening and drying yellowish or reddish brown, near base faded yellow or straw colored, the color finely stippled and obscurely streaked longitudinally with brown; ripens midseason.	Fair quality date. California.
Bayjoo*	Badjou*	Tunisia 1905	——	Dry; small; oblong-elliptical, a straighter and more pronounced taper from greatest diameter to base and apex; yellow, ripening dull buff, darkening a little with age; ripens early.	Mediocre date. Arizona.
Bent el Fgee*	——	Algeria? 1908?	——	Soft; large; oblong; dull amber curing to dark reddish brown; ripens midseason.	Inferior date; fruit are subject to severe checking in long transverse scars from base to apex. Arizona.
Bent el Maroo*	——	Algeria? 1908?	——	Soft; large; oblong; yellow, ripening dull amber, curing dark reddish brown; ripens early.	Fruit sour easily and are quite susceptible to checking in fine, short, irregular scars. Arizona.
Bentamoda	Betamoda, Bartomoda, Bartamuda	Sudan 1914 (Mason)	Similar to Deglet Noor but differs in less vigorous growth; shorter spine area with fewer, shorter, and weaker spines; shorter pinnae; and solid bands of fiber, 3–4 cm wide across leaf bases.	Semidry to dry; medium to large; narrowly oblong-elliptical; dull red with some orange-yellow background, ripening amber, curing darker brown; flesh firm with thin layer of rag; flavor good; ripens late.	Reputed to be a superb dessert fruit in Sudan but has not attained this quality in U.S.; fruit are similar to those of Deglet Noor but are slightly drier, more susceptible to checking, and less flavorful; yield is lower than that of Deglet Noor. California. In Thermal, Brawley, and Tempe germplasm collections.
Besser Haloo	Bisra Haloua, Busr Hulu	Tunisia 1904 (Kearney)	Short, stiff leaves and pinnae forming a small, compact, bristly crown; short fruitstalks.	Dry; small; oblong-elliptical or somewhat obovate; yellow, ripening and curing light brown or straw colored with light to moderate bloom; flesh firm, becoming dry and hard with age; flavor agreeably sweet; ripens midseason.	Reputed to be highly productive and very salt resistant in Tunisia but productivity has been lower than other dry dates in the U.S.; fruit are subject to little damage, mostly checking, from rain. Arizona, California.
"Blue Thoory"	——	Algeria 1913 (Johnson)	Leaves with heavy bloom lending bluish cast (hence name); fruit with small perianth set in a depression.	Dry; medium to large; oblong with rounded apex; flesh tough, fibrous.	Palm has little resemblance to true Thoory; fruit are about same size and shape as those of Thoory but skin is more wrinkled, flesh tough and fibrous, and flavor distinctly inferior. California.
"Boo Affar"	——	Tunisia 1905	——	Semidry; oblong with rounded apex, slightly wider above middle; yellow, ripening amber, curing duller, darker shade of brown; ripens early.	Inferior date; fruit are subject to considerable damage from rain; true Boo Affar not cultivated in the U.S. Arizona.
Boo Halas*	——	Algeria 1904	——	Dry; small; oblong; yellow, ripening and curing light grayish brown with softer portions dull amber; ripens early.	Mediocre date. California.
"Deglet Barca-1"	——	Tunisia 1905 (Kearney)	——	Semidry; medium to large; oblong-elliptical; red, ripening and curing nearly black; ripens early.	Inferior date; fruit of true Deglet Barca, from Tunisia and not cultivated in the U.S., are soft, round, and nearly black. Arizona.

Varietal name and meaning	Synonyms	Origin, date of introduction (and introducer)	Distinguishing characters	Fruit	Notes
"Deglet Barca-2"	——	Tunisia 1905 (Kearney)	——	Soft; large; oblong-elliptical; yellow, ripening dull amber, curing reddish brown; ripens late.	Mediocre date; fruit are subject to less conspicuous checking than those of "Deglet Barca-1" but tend to sour easily; fruit of true Deglet Barca, from Tunisia and not cultivated in the U.S., are soft, round, and nearly black. Arizona.
Deglet Beida *white date* (Anonymous 1920)	Degla Beida, Daqlah Bayda	Algeria 1904	Medium-heavy trunk; long leaves with moderate curvature increasing distally; wide, green with a little reddish brown leaf bases narrowing rapidly above fiber line; spine are about one-sixth of blade length; numerous, slender to medium-heavy, mostly paired spines 8–14 cm long lacking necks; medium-long and wide, somewhat drooping pinnae in groups of 2, 3, or 4; fruit differ from that of any other imported variety in their very light color, relatively smooth skin, and oblique base.	Dry; medium to large; oblong or narrowly oblong, base obliquely flattened, widest usually a little below middle with a slight taper to the bluntly rounded apex; yellow, ripening and curing very light pale brown or buff; ripens early.	Principal dry date of Algeria and said to be quite salt resistant; fruit subject to occasional checking; fruit about the size of those of Thoory but are drier, harder, and of inferior quality. Arizona. In Thermal, Brawley, and Tempe germplasm collections.
Dishtari* *the bride's tree*	——	Pakistan (Baluchistan) 1902	——	Soft; small to medium; oval; yellow, ripening amber, curing dark reddish brown with moderate bloom lending a purplish cast; ripens early.	Second-rate date; fruit are subject to only slight to moderate damage from rain. Arizona.
Doonga*	——	Tunisia 1905	——	Dry; small; oblong-ovate; yellow, ripening dull amber, curing dull chocolate brown; ripens early.	Second-class date. Arizona.
Frahee* *of little joy*	Freyeh*	Egypt 1905	——	Soft; medium to large; oblong with somewhat greater diameter above middle; yellow, ripening deep dull brownish red, curing nearly black; ripens midseason.	Inferior date; fruit do not cure well. Arizona.
Fursi *the Persian* (Popenoe 1913b) or possibly *of furs* (Dowson 1939)	Firse, Farsi, Farisi	Iraq 1913 (Popenoe)	Medium-heavy trunk; medium-long leaves with moderate curvature near tip; medium-wide, maroon leaf bases; medium to stout, paired spines 10–16 cm long with neck 2–3 cm long; short, wide, stiff pinnae mostly in groups of 2; somewhat similar to Dayri but differs in the deeper, solid red *khalal* fruit and the maroon color at base of older leaves.	Semidry; medium; oblong with rounded apex; red, ripening dull reddish brown, softer fruit curing nearly black, drier fruit changing less, with moderate bloom lending purplish cast; flesh soft to firm but not tough, seldom very hard; ripens late.	Well known and widely distributed around Basra, Iraq, but nowhere abundant; one of the best black dates of Basra but fruit are very susceptible to rot in humid weather in the U.S.; date is borderline between soft and semidry; softer fruit is attractive and of good quality. California.
Gaggar*	——	Egypt 1904	——	Semidry; medium; oblong; orange-yellow, sometimes with fine reddish brown stippling, ripening dull amber, curing deep reddish brown; ripens midseason.	Mediocre date; fruit are subject to severe damage from rain in Arizona, mainly checking in characteristically small, short scars, many mere dots. Arizona.
"Gale's Palm"	——	Algeria? Egypt? 1890	——	Soft; oblong; red, ripening and curing dull dark brown; ripens midseason.	Mediocre date. California.
Gantar *the hundred-weight,* probably referring to its yield (Popenoe 1913b)	Guntar, Qintar	Iraq 1913 (Popenoe)	Narrow, yellowish green leaf bases; stiff pinnae with considerable crisscrossing in mid and proximal portions of blade; long leaves with moderate curvature; spine area about one-fifth or less of blade; and medium-stout, more or less paired spines 10–16 cm long on necks 1–2 cm long.	Soft; small; oblong-oval; yellow, ripening amber, curing deep reddish brown; ripens midseason.	Considered one of the better varieties of southern Iraq; excellent date but small size is a drawback; fruit are subject to only slight damage during humid weather. California.
Gasb Haloo*	——	Tunisia 1905	——	Dry; medium; narrowly oblong; yellow, ripening and curing dull reddish brown.	Second-rate date; fruit are subject at times to severe checking. Arizona.
Gasby* *of the reed*	——	Tunisia 1905	——	Soft; large; narrowly oblong, sometimes slightly curved to one side; yellow, ripening dull brown, curing nearly black with moderate to heavy bloom lending purplish cast; ripens very early.	Puffy date of mediocre quality; fruit are subject to considerable losses from drop but rain damage has not been serious. Arizona.

Varietal name and meaning	Synonyms	Origin, date of introduction (and introducer)	Distinguishing characters	Fruit	Notes
Gondeila *the little lamp*	Gondaila, Gondela, Gondila, Gundila	Sudan 1922 (Mason)	——	Dry; large; oblong-ovate; yellow, ripening light brown or amber, becoming somewhat duller and darker with age; ripens late.	Highly rated in Sudan and a date of some merit; rather tough flesh is drawback. California.
"Goondy"	"Goondee," "Goondie"	Tunisia 1905	——	Dry; small; ovate; yellow, ripening amber, curing clay or russet colored; ripens late.	Inferior date; fruit subject to moderate damage from rain; fruit of true Goondy (not cultivated in the U.S.) are soft. Arizona.
"Gush" *a male palm*	——	Pakistan (Baluchistan) 1902	Medium-heavy trunk; long, stiff leaves with slight to moderate curvature near tip; medium-wide, green leaf bases with slight maroon and sparse scurf on edges; long petioles with unusually sharp margins; spine area one-fifth of blade length; 10-14, solitary spines 10-15 cm long on slender, weak necks 1 cm long; medium-long and wide, drooping pinnae.	Soft; medium; oblong, tapering slightly to bluntly pointed apex; red, ripening and curing deep maroon, nearly black with moderate bloom lending purplish cast; ripens early.	Imported as staminate variety but proved to be pistillate; the two abortive carpels often develop after pollination, attaining 1 cm long or more and persisting until fruit matures; fair quality date; fruit are subject to only slight damage from rain. Arizona.
Halooa-1* *sweetmeat*	Halloua*, Heloua*	Algeria 1900	Palm similar to Halooa-2* but differs in more slender trunk, more open crown, shorter leaves with narrower leaf bases, and shorter, narrower, and stiffer pinnae with greater apical divergence.	Dry; small to medium; oblong-obovate; yellow, ripening amber, curing dull reddish brown; ripens early.	Inferior quality date but reputed to be highly esteemed in Algeria for purported value of fruit in preparing an aphrodisiac; fruit are subject to only slight damage from rain. Arizona.
Halooa-2* *sweetmeat*	Halloua*, Heloua*	Algeria 1904	Palm similar to Halooa-1* but differs in heavier trunk; denser crown; longer leaves with broader leaf bases; and longer, broader, and more drooping pinnae with smaller apical divergence.	Semidry; small to medium; oblong-ovate; yellow with greenish tinge, ripening dull amber, curing dull dark brown; ripens early.	Inferior quality date but reputed to be highly esteemed in Algeria for purported value of fruit in preparing an aphrodisiac; fruit are subject to moderate damage from rain. Arizona.
Halooa Bayda* *white sweetmeat* (Kearney 1906)	——	Tunisia 1905	——	Semidry; small; oval; yellow, ripening and curing dull brown; ripens late.	Poor quality date; seed are large in proportion to fruit; fruit are subject to only slight damage from rain. Arizona.
Hamra Bischry	Hamra Bechry, Hamra Bishri	Algeria 1912 (Johnson)	——	Soft; large; oblong, tapering a little above middle to bluntly pointed apex; dull red over more or less prominent yellow background, ripening dull dark brown and becoming nearly black with heavy bloom lending purplish cast; flesh soft, smooth, with little rag; flavor suggestive of molasses; ripens late.	Fair quality date, but the fruit has not been handled or observed sufficiently to determine curing and storage capabilities; fruit are subject to some checking. California.
Hasan Efendi*	Hasan Effendi*, Hussein Effendi*	Iraq 1908	——	Semidry; medium; obovate.	Rare and commercially unimportant in Iraq, somewhat similar to Maktoom; probably named for a man, perhaps the owner of the palm (Popenoe 1913b); palm and fruit are comparable to those of Zahidi but fruit are slightly larger, a little softer, and somewhat later in ripening. California.
Jafari *of Ja'far*	Jaafari	Iraq 1929 (Nixon)	——	Soft; medium; oblong-oval; deep red, ripening nearly black with light bloom; ripens midseason.	Rated as one of the better varieties in Iraq; tendency for fruit to drop after ripening is drawback. California, Texas.
Kaiby*	——	Egypt 1905	Palm is similar to Saidy but differs in less harsh leaves and longer, narrower, and more evenly arranged pinnae.	Semidry; large; oblong to oblong-oval; yellow tinged with orange, ripening amber, curing chestnut sometimes light brown or buff near base; ripens early to midseason.	Date of some merit, but fruit are subject to considerable spoilage from humid weather. Arizona, California.

Varietal name and meaning	Synonyms	Origin, date of introduction (and introducer)	Distinguishing characters	Fruit	Notes
Kalara*	Kularu*	Pakistan (Baluch-istan) 1902	Medium-heavy trunk; medium-long leaves with moderate curvature increasing distally; medium-wide, green leaf bases with maroon on edges; spine area about one-tenth of blade length; 14–18 mostly solitary spines 8–12 cm long; medium-long and wide, stiff pinnae seldom drooping but occasionally bending or breaking abruptly.	Soft; small; oblong-elliptical, widest usually a little above middle; yellow, ripening amber, curing deep reddish brown, often nearly black at base; flesh caramel-like with nearly no rag; flavor rather delicate, pleasing; ripens early.	Fruit cure well, seem less suscep-tible to fruit drop than most other soft dates, and are subject to very little damage from rain or humid weather. California.
Karooy*	——	Tunisia 1905	——	Semidry; large; oblong-ovate to oblong-oval; yellow, ripening dull brown with little further change; flesh firm with some tough rag; ripens midseason.	Very inferior date. Arizona.
Khadrawy of Baghdad*	——	Iraq 1929 (Nixon)	Vigorous palm of rapid growth; pinnae diverging rather sharply, some crisscrossing even near the tip, proximal pinnae bending abruptly lending a tangled appearance to the center of the crown; fruit are similar to those of true Khadrawy but differ in larger size with more green tinge in *khalal* stage and nearly double the yield.	Soft; medium; oblong-elliptical to oblong-ovate; yellow tinged with green, ripening amber, curing reddish brown; flesh melting, caramel-like, rag slight or lacking; ripens late.	Highly regarded and extensively grown in northern Iraq; palm differs from true Khadrawy in more vigorous, rapid growth and longer leaves with larger, stouter, and more numerous spines and sharply diverging pinnae lending blade a rough, tangled, asymmetrical appearance. California, Texas.
Khalt Kebeer*	——	Tunisia 1905	——	Large; soft; oblong-ovate; yellow, ripening brown, curing dull brown, often somewhat reddish; flesh spongy with considerable rag.	Date of no apparent merit. Arizona.
Khatuni* *of the lady* (Dowson 1939)	Khatooni*	Iraq 1929 (Nixon)	——	Soft; medium; oblong-ovate; red, ripens nearly black; ripens late.	Fair quality date; fruit have ten-dency to shrivel and are subject to moderate checking. California, Texas.
Koroch* *seedling*	Kuroch*, Kuruch*, Korroo*, Bagum Jurghi*, Rogani*	Pakistan (Baluch-istan) 1902	Slender trunk; moderately long, very stiff leaves; medium-wide, yellowish green leaf bases; spine area one-fifth of blade length; 20–26 slender to medium-heavy spines 12–18 cm long; stiff, more or less evenly spaced, only slightly drooping.	Soft; small to medium; ovate; red, ripening deep brown, curing nearly black with heavy bloom lending purplish cast; flesh stringy with much rag; ripens very early.	Mediocre date; fruit are subject to pronounced shrinkage and moder-ate spoilage from rain. Arizona, California, Texas.
Kseba* *the profitable* (Popenoe 1913b); *the little reed*	Kasbeh*, Kesba*, Kesseba*, Kessebi*, Ksebba*	Algeria 1904	Rather heavy trunk; wide, yellowish green leaf bases with some maroon near fiber; spine area one-fourth of blade length; 26–36 mostly paired spines 10–18 cm long with heavy neck 1–3 cm long; long pinnae slightly drooping with age.	Soft; medium; oblong-oval; yellow, ripening amber, becoming deep dull reddish brown; ripens early to midseason.	Mediocre date; fruit are subject to serious checking, souring, and rot from humid weather. Arizona.
Kush Sheham* *the pulpy* (Popenoe 1913b)	Qush Sheham*, Qush Shahm*	Oman 1913	——	Soft; medium; oblong-elliptical; yel-low, ripening dull amber, curing deep reddish brown; ripens very early.	Fair quality date but fruit are subject to severe souring and drop in humid weather. California.
"Lagoo-1" perhaps *the distorted mouth* (Popenoe 1913b)	"Lagou," "Laqu"	Tunisia 1905	Heavy trunk; long, stiff leaves; wide, green leaf bases; wide pinnae.	Dry; large; oblong-elliptical; dull red-dish yellow with some greenish yellow, ripening chestnut, becoming some-what deeper with age; ripens early.	Very inferior date; fruit of true Lagoo (not cultivated in the U.S.) are soft. Arizona.
"Lagoo-2" perhaps *the distorted mouth* (Popenoe 1913b)	"Lagou," "Laqu"	Tunisia 1905	Slender trunk; short, stiff leaves, narrow, green leaf bases, short, narrow pinnae.	Dry; large; oblong-ovate; dull yel-low, ripening dull yellowish brown, becoming slightly darker with age; ripens early.	Very inferior date; fruit of true Lagoo (not cultivated in the U.S.) are soft. Arizona.
"Leonard's Unknown"	——	? (Johnson?)	——	Soft; large.	Was known only from three palms in abandoned planting near Thermal, California; attractive date of Medjool type but fruit are subject to severe checking. California.
Lookzy*	Lookzee*, Rhazee*	Algeria 1904	Unusual and distinctive reddish orange fruitstalks.	Soft; small; narrowly oblong; yellow, ripening amber, curing dark brown; ripens early.	Inferior date; fruit are subject to moderate spoilage from rain or humid weather. Arizona.

Varietal name and meaning	Synonyms	Origin, date of introduction (and introducer)	Distinguishing characters	Fruit	Notes
Makelet el Leef* *the female eater of the palm fiber*	Makilat al-Lif*	Algeria 1904	——	Soft; large; oblong-elliptical or slightly oblong-obovate; yellow, ripening amber, becoming darker but not curing well; ripens early.	Second-rate date; fruit are subject to considerable checking. Arizona.
Medina* name of city in Saudi Arabia	——	Sudan 1927	——	Soft; large; oblong; yellow, ripening dull amber, curing deep reddish brown with blistered areas dull earthen brown; ripens early.	Reputed to be a rare and excellent date of some merit, but fruit do not cure well and are subject to severe checking. California.
"Middleton's Unknown"	——	Algeria?, Iraq? 1913 (Popenoe)	Palm similar to Rhars but differs in larger rachis angles of pinnae, neck of spine only half as long, and narrower leaf bases with old ones having more maroon color on lower portions.	Soft; large; narrowly oblong-obovate; yellow, sometimes with faint red tinge near base; ripening light amber, curing reddish brown; flesh melting with some rag; ripens early.	Fruit similar to those of Rhars, but are smaller and more pointed, sometimes have faint red tinge near base, are lighter-colored when ripe, have thinner skin and more rag, and are less susceptible to damage from rain. California.
Mokh Begry *the brain of an ox* (Kearney 1906)	Moukh Begri	Tunisia 1905 (Kearney)	Medium-heavy trunk; leaves with slight to moderate curvature increasing distally; narrow, slightly yellowish green leaf bases with old ones having a little maroon mottling near fiber; spine area one-fifth of blade length; numerous, stout, mostly paired spines 8–10 cm long lacking a neck; mostly stiff pinnae in groups of 2 and 3, occasionally drooping slightly.	Soft; small to medium; oblong-blocky with base and apex more or less flattened; yellow with more or less fine brown stippling, ripening amber; curing reddish brown; ripens late.	Reputed to be extremely rare and much esteemed in Tunisia; fruit are not subject to damage from rain but small size and odd shape are drawbacks. California.
"Mozaty"	——	Pakistan (Baluchistan) 1902	Medium-heavy trunk; long leaves, moderate to pronounced curvature in distal half; green, slightly glaucous leaf bases with moderate scurf on edges; spine area one-tenth of blade length; few, solitary spines lending an open center to crown; slight to moderately drooping pinnae in proximal part of blade.	Soft; medium; oblong-elliptical, widest a little above middle; yellow, ripening amber, curing bay or nearly black with light bloom; ripens early.	Second-rate date; fruit are subject to only slight spoilage during humid weather and some checking; fruit of true Mozaty (not cultivated in the U.S.) have different shape and color and are superior in quality. Arizona.
M'Sillia*	——	Algeria 1908 (Johnson)	——	Soft; medium; oblong, tapering somewhat above middle to bluntly rounded apex, frequently slightly constricted between middle and somewhat oblique base; yellow or slightly orange-yellow, ripening amber, curing reddish brown.	Flavor of fruit is suggestive of Zahidi. California.
"Nagal" *the bastard*	"Naghl", "Naaghal"	Oman (Muscat) 1902	——	Soft; medium; black; ripens very late.	One of latest ripening of all imported varieties; fruit subject to little damage from rain or humid weather; palm and fruit are similar to those of Khisab and variety may not be distinct; fruit of true Nagal (not cultivated in the U.S.) are amber and ripen very early. Arizona.
Najl al Pasha* *son of the pasha*	Nagl el Basha*, Nazl el Basha*	Egypt 1901	Palm and fruit are similar to those of Hayany but differ in leaves with shorter pinnae with only slight to moderate drooping and greater apical divergence; fewer spines; wider leaf bases; and fruit are lighter red in *khalal* stage and not nearly so black when ripe.	Soft; large; oblong-elliptical; red with some orange-yellow background, ripening and curing very dark reddish brown; ripens early.	Fruit subject to serious spoilage from humid weather. Arizona.
Nakhelet Feraoon*	Nakhlat Fir'aun*	Algeria 1904	——	Soft; large; oblong-oval; yellow with minute brownish stippling, ripening amber, curing slightly deeper shade of brown; ripens early.	Second-rate date; fruit are subject to considerable spoilage during humid weather; fruit usually souring before curing completely. Arizona.
Nakleh el Pasha* *the pasha's palm* (Popenoe 1913b)	Neklet el Pasha*, Nakhlet al Pasha*	Egypt 1890	——	Soft; small; oblong-obovate; dull red, ripening dark brown and becoming nearly black; ripens midseason.	Mediocre date. California.

Varietal name and meaning	Synonyms	Origin, date of introduction (and introducer)	Distinguishing characters	Fruit	Notes
Nakleh Zianeh* *the beautiful palm*	Nakleh Zian*	Algeria 1913	——	Soft; medium; oblong-elliptical; yellow, ripening amber, curing dark brown; ripens early.	Mediocre date; fruit are subject to considerable spoilage during humid weather. California.
"Near Khalasa"	——	? 1913 (Popenoe)	Palm similar to Khalasa but differs in less glaucous, deeper green leaves with more drooping pinnae and longer terminal pinnae.	Soft; medium to large; oblong, tapering slightly from middle to rounded apex; yellow, ripening amber, curing reddish brown, translucent with light to moderate bloom; ripens early.	Name was given because fruit have some resemblance to those of Khalasa but are larger and longer; attractive date but fruit are subject to heavy losses in humid weather. California.
Nesheen*	——	Algeria 1904	——	Soft; small; oblong-elliptical; yellow, ripening dull brown, curing dark reddish brown; ripens early.	Mediocre date; fruit are subject to only little damage from rain. Arizona.
Okht Fteemy-1* *sister of Fteemy* (Kearney 1906)	——	Tunisia 1905	Slender trunk; leaves with slight curvature in distal half; medium to medium-broad, green leaf bases with slight maroon mottling in center of old ones; spine area one-third of blade; numerous spines 12–16 cm long in groups of 2 and 3 and set at widely divergent angles; short, wide, stiff pinnae.	Soft; medium; oblong-elliptical, widest usually above middle; yellow, usually with reddish tinge, ripening amber, curing deep reddish brown; ripens early.	Fair quality date; fruit are subject to slight to moderate checking and splitting from rain or high humidity. California.
Okht Fteemy-2* *sister of Fteemy* (Kearney 1906)	——	Tunisia 1905	Palm is similar to Okht Fteemy-1 but differs in leaves with more curvature in distal half; moderate to heavy scurf on edges of leaf bases; spine area one-fifth of blade; and medium number of not-so-heavy spines.	Semidry; medium; oblong-elliptical; yellow with red tinge, ripening amber, curing deep reddish brown; ripens midseason.	Fair quality date; fruit are subject to checking and blacknose. Arizona.
Oogbales*	——	Algeria 1904	——	Dry; medium; oblong; yellow, ripening and curing dull dark earthen brown; ripens early.	Inferior date; fruit are subject to severe fruit drop. Arizona.
Remta*	Remtsa*	Tunisia 1905	Stiff leaves; stiff, appressed, crowded pinnae; and long, stout spines with a 3–4 cm long neck.	Dry; medium; oblong, tapering somewhat to bluntly pointed apex; yellow, ripening and curing yellowish or dull brown; ripens early.	Fair quality date; fruit are subject to only slight damage from rain or humid weather. Arizona.
Retbet Abdala* *Abdallah's fresh date*	——	Algeria 1904	——	Soft; small; obovate to oblong-obovate; yellow, ripening light brown, curing dull dark brown; ripens early.	Inferior date; fruit are subject to only slight spoilage from rain or humid weather. Arizona.
Retbet Hafsia* *Hafsia's fresh date*	Rutbut Hafsah*	Algeria 1904	——	Dry; medium; oblong; yellow, ripening and curing dull amber brown; flavor good; ripens midseason.	Fruit are subject to only slight spoilage during humid weather. Arizona.
Retbet Regaya-1* *Regaya's fresh date*	Retbet Regaia*	Algeria 1904	——	Semidry; large; oblong-elliptical; yellow, ripening amber, curing reddish brown; ripens midseason.	Second-rate date. Arizona.
Retbet Regaya-2* *Regaya's fresh date*	Retbet Regaia*	Algeria 1904	——	Soft; large; narrowly oblong; red, ripening and curing nearly black; ripens early.	Inferior date. Arizona.
Roghm Gazal* *muzzle of the gazelle*	Ruhm al Ghazel*	Egypt 1905	——	Dry; medium; oblong, tapering from middle or slightly above to bluntly pointed apex; yellow, ripening and curing variously from light grayish brown to dark reddish brown; ripens midseason.	Fair quality date; in Arizona once known erroneously as Frahee. Arizona, California.
"Russel's No. 17"	Hasawi*	Iraq 1913 (Popenoe)	——	Soft; medium to large; oblong, tapering to bluntly pointed apex; yellow, ripening amber, curing only slightly darker; flesh melting; ripens early.	California.
"Russel's No. 23"	Asabi al-Arus*	Iraq 1913 (Popenoe)	——	Soft; large; oblong, tapering slightly from middle to apex; yellow, ripening amber, curing reddish brown; flesh melting; ripens late.	California.
"Russel's No. 32"	Swaidan*	Iraq 1913 (Popenoe)	——	Soft; large; oblong; yellow, ripening amber, curing reddish brown; flesh slightly "grainy"; ripens early.	California

Varietal name and meaning	Synonyms	Origin, date of introduction (and introducer)	Distinguishing characters	Fruit	Notes
"Russel's No. 40"	Baljani*	Iraq 1913 (Popenoe)	——	Soft; medium to large; oblong-elliptical; yellow, ripening amber, curing reddish brown; flesh caramel-like; ripens early.	California.
"Russel's No. 48"	Nabaity*	Iraq 1913 (Popenoe)	——	Soft; medium; oblong; yellow, ripening amber, curing reddish brown; flesh with considerable rag around seed; ripens early.	California.
"Russel's No. 50"	Khasawi al-Baghl*	Iraq 1913 (Popenoe)	——	Soft; large; oblong-elliptical; yellow, ripening amber, curing reddish brown; flesh with very little rag; ripens late.	California.
Safraia-1* *yellow one*	Safrayah*	Algeria 1900	——	Dry; medium; narrowly oblong or oblong-elliptical; pale yellow, ripening and curing dull faded yellow or straw color with fine brown stippling; flesh thin, papery but not tough; flavor pleasing; ripens early.	Inferior date but easily the best of the three under the name "Safraia"; fruit are subject to considerable checking. Arizona.
Safraia-2* *yellow one*	Safrayah*	Algeria 1900	——	Semidry; medium; broadest near oblique base, tapering to bluntly pointed apex; yellow with faint red tinge near base, ripening dull amber, curing reddish brown; flesh firm; flavor insipidly sweet; ripens midseason.	Inferior date. Arizona.
Safraia-3* *yellow one*	Safrayah*	Algeria 1900	——	Semidry; small; oblong-elliptical; yellow, ripening and curing light brown with occasional straw-colored dry areas; flesh somewhat mealy; flavor sweet but rather flat; ripens midseason.	Inferior date; fruit a little softer than those of Safraia-2. Arizona.
Sba Aroosa* *bride's fingers (Kearney 1906)*	——	Tunisia 1905	——	Semidry; large; tapering from near pronounced oblique base to pointed apex; yellow, ripening light brown, curing reddish brown with occasional straw-colored areas at base; ripens early.	Inferior date. California.
Sebaa Loosif*	Sayba Loosif*	Algeria? 1904 or Pakistan? (Baluchistan) 1902	——	Soft; medium; narrowly oblong; yellow, ripening and curing nearly black with moderate bloom lending purplish cast; ripens early.	Very inferior date. Arizona.
Shitwi Asfar* *yellow winter date*	——	Iraq 1908	Long, slightly arched leaves arranged in vertical ranks; broad, glaucous green leaf bases with maroon edges; narrow, stiff pinnae.	Soft; small; nearly round; yellow, ripening amber, curing deep reddish brown; ripens late.	Second-rate date. California.
Shukkar *sugar (Dowson 1939)*	Shukar	Iraq 1913 (Popenoe)	Palm similar to Khadrawy but differs in longer leaves lacking a definite "fishtail" at tip and duller green leaf bases with maroon mottling on edges with age.	Soft; medium; oblong-elliptical; yellow, ripening amber, curing reddish brown; ripens early.	One of rarer varieties of Iraq; date of Sayer type but fruit are smaller. California.
Sokria* *sugary*	——	Algeria 1904	——	Dry; medium; narrowly oblong; yellow, ripening and curing buff or light brown; ripens midseason.	Mediocre date. California.
"Sukkar Nabat" *sugar candy*	"Shuccar Nabat"	Iraq 1913	——	Soft; medium; oblong, oblique base; yellow, ripening amber, curing dark reddish brown; ripens early.	Date of indifferent quality; fruit are subject to considerable souring during humid weather; fruit of true Sukkar Nabat (not cultivated in U.S.) are dry, small, and yellow. California.
"Sukkari" *of candy (Dowson 1939)*	"Sukeri"	Iraq 1902	——	Semidry; medium; oblong-elliptical; red, ripening and curing nearly black; ripens early.	Fair quality date; fruit are subject to severe souring and checking during humid weather; fruit of true Sukkari (not cultivated in the U.S.) are soft, medium size, and amber. Arizona.

Varietal name and meaning	Synonyms	Origin, date of introduction (and introducer)	Distinguishing characters	Fruit	Notes
Tafazweent*	——	Algeria 1904	——	Semidry; large; narrowly oblong-elliptical; yellow with various amounts of red, ripening and curing dull brown on drier parts and much darker on soft portions; ripens midseason.	Inferior date; fruit are subject to moderate splitting and checking from rain. Arizona.
Takadet*	——	Algeria 1904	Short, moderately arched leaves; green leaf bases maroon abaxially with age; spine area one-fourth of blade; spines in groups of 2; narrow pinnae drooping slightly with age.	Soft; medium; broadly oblong with rounded apex; yellow with fine brown stippling, ripening and curing dull dark brown; ripens early.	Fair quality date; fruit are subject to souring and checking from rain. Arizona.
Takermest	Takarmist, Takermoust	Algeria 1912 (Johnson)	Slender trunk; short, slightly curved leaves; narrow to medium-wide, yellowish green leaf bases, old ones maroon abaxially with faded tan-colored areas on edges; spine area one-fifth of blade; short, slender spines arranged mostly in twos; narrow, slightly to moderately drooping.	Semidry; medium-large or larger; round or nearly so; yellow, ripening and curing dull brown; ripens midseason.	Mediocre date; fruit blister considerably in curing. California.
Tamzoohart*	——	Algeria? 1904 or Pakistan? (Baluchistan) 1902	——	Soft; small; oblong-oval; yellow, ripening amber, curing dark reddish brown; ripens early.	Fair quality date; fruit subject to only slight damage during humid weather. Arizona.
"Tantaboosht"	"Tanta-boucht", "Tenta-boosh"	? ?	——	Semidry; medium; oblong; yellow, ripening dull amber, curing somewhat darker; ripens late.	Mediocre date; fruit are subject to considerable irregular checking; fruit of true Tantaboosht (not cultivated in the U.S.) are soft, spherical, black, and of inferior quality. Arizona.
Taoorkhet*	Toorekhet*	Algeria 1904	——	Semidry; large; narrowly oblong, apex turned a little to one side; dull red over dull orange background, ripening and curing deep maroon or nearly black; ripens early.	Inferior date; fruit are subject to moderate rot and severe checking in humid weather. Arizona.
Taremoont*	——	Algeria 1904	——	Soft; medium to large; oblong with rounded apex and slightly greater diameter above middle; dull orange red, ripening and curing nearly black; ripens midseason.	Inferior date; fruit are subject to severe checking and souring in humid weather. Arizona.

Varietal name and meaning	Synonyms	Origin, date of introduction (and introducer)	Distinguishing characters	Fruit	Notes
Taty*	——	Algeria 1904	Crown of leaves suggestive of Hayany but differs in denser center with more asymmetrically arranged pinnae; green leaf bases with faint yellowish red tinge, a little maroon mottling with age and moderate scurf on edges; spine area one-eighth of blade; and few to medium number of spines with 2–4 cm long neck.	Soft; medium; oblong-oval to oblong-obovate; yellow, ripening amber, curing dark reddish brown; ripens early.	Good date in favorable seasons but fruit ferment easily and are subject to severe checking during humid weather. Arizona.
Taurarket*	——	Algeria 1900	Slender trunk; glaucous green, moderately and uniformly curved leaves; green leaf bases with a little maroon abaxially; pinnae drooping slightly.	Soft; large; oblong, tapering somewhat toward apex; yellow, ripening amber, curing dark reddish brown; ripens early.	Fairly good quality date; fruit are subject to only slight damage from rain. Arizona.
Tenaseen	Tenacine, Tennessin, Tanasin, Tanessin, Tenassine, Tinissin	Algeria 1900	Moderately arched leaves; green, slightly glaucous leaf bases with a little maroon on edges with age; spine area one-eighth of blade; few, short, slender, single spines; narrow, moderately to pronounced drooping pinnae.	Soft, medium; oblong-elliptical; yellow, ripening and curing nearly black, especially at base; ripens early.	Puffy date of very little merit; fruit are subject to only slight checking from rain. Arizona.
"Timjooert"	——	Algeria 1900–1906	——	Soft; medium; oblong-elliptical; red (perianth yellow), ripening reddish brown, curing dark purplish brown, nearly black; ripens midseason.	Mediocre date; fruit are somewhat similar to those of Timjooert-Red* but are only slightly susceptible to checking; true Timjooert is not cultivated in the U.S. California.
Timjooert-Red* uncertain, perhaps *earring date*	Timjouert*, Timjuhart*, Timdje-houert*, Tamedjo-hert*	Algeria 1900–1906	Medium-heavy trunk; long leaves, moderate curvature increasing distally; leaf bases with a little maroon on edges; spine area one-seventh of blade; medium number of short, medium-heavy spines mostly in twos, without neck; medium to short pinnae slightly drooping with age.	Soft; medium; oblong-ovate; red, ripening dull amber, curing nearly black; ripens early.	Very similar to true Timjooert from Algeria (not cultivated in the U.S.) and possibly identical with it; fruit are extremely susceptible to checking and blacknose. Arizona.
Timjooert-Yellow* uncertain, perhaps *earring date*	Timjouert*, Timjuhart*, Timdje-houert*, Tamedjo-hert*	Algeria 1900–1906	Slender trunk; leaves with slight to moderate curvature; green, narrow leaf bases with slight scurf on edges; spine area one-sixth of blade; short, medium stout spines mostly in twos with 2–3 cm long neck; medium to short pinnae drooping slightly with age.	Soft; medium; narrowly oblong, widest a little above middle, one side slightly concave; yellow, ripening amber, curing reddish brown; ripens early.	Fair quality date; fruit are not very susceptible to checking or other damage from high humidity; large size of seed is drawback. Arizona.
Toojat*	——	Algeria? 1904 or Pakistan? (Baluchistan) 1902	——	Dry; large; obovate-oblong, tapering rather abruptly from greatest diameter to bluntly pointed apex; yellow, ripening and curing dull amber on softer portions to light chocolate or faded brown on drier portions; ripens early.	Very inferior date; fruit are subject to serious checking, splitting, and rot during humid weather. Arizona.
"Towadant"	"Taoudant"	Tunisia 1905	——	Dry; small; oblong-oval; yellow, ripening light brown, curing duller shades; ripens midseason.	Inferior date; fruit are subject to considerable losses from drop; fruit of true Towadant (not cultivated in the U.S.) are soft, large, long, yellow, and of good flavor. Arizona.
"Tronja" *the citron* (Popenoe 1913b)	"Troundja", "Turunja"	Tunisia 1905	——	Semidry; small; oblong-elliptical; yellow, ripening dull brown, changing little in curing except for fading on drier portions; ripens midseason.	Inferior quality date; fruit of true Tronja (not cultivated in the U.S.) are soft, large, and globular with thick flesh. Arizona.

Varietal name and meaning	Synonyms	Origin, date of introduction (and introducer)	Distinguishing characters	Fruit	Notes
Zerza*	——	Algeria 1904	——	Dry; medium; oblong, apex rounded, obtusely pointed on one side; yellow, ripening and curing dull amber on softer portions and lighter faded brown on drier portions; ripens midseason.	Fair quality date; fruit are subject to only occasional checking during humid weather. Arizona.
Zoozia*	——	Algeria 1904	——	Dry; small to medium; oblong-rounded; yellow, ripening and curing light brown; ripens midseason.	Second-rate date; possibly same as Zouza from Tunisia. Arizona.
Zrai* perhaps *little produce*	——	Tunisia 1905	——	Soft; medium; oblong-elliptical, oblique base usually pronounced with perianth on one side; yellow, ripening amber, curing reddish brown; ripens midseason.	Mediocre date; fruit are subject to only slight to moderate damage and little checking from rain. Arizona.

LITERATURE CITED

Abdul-Baki, A., S. Aslan, E. H. Beardsley, S. Cobb, and M. Shannon. 1998. Soil, water, and nutritional management of date orchards in Coachella Valley and Bard. Washington DC: USDA.

Albert, D. W., and R. H. Hilgeman. 1935. Date growing in Arizona. Tempe AZ: Ariz. Agr. Expt. Sta. Bull. 149.

Aldrich, W. W. 1942. Some effects of soil moisture deficiency upon Deglet Noor fruit. Date Growers' Inst. Rep. 19:7–10.

Aldrich, W. W., and C. L. Crawford. 1941. Second report upon cold storage of date pollen. Date Growers' Inst. Rep. 18:5.

Aldrich, W. W., C. L. Crawford, R. W. Nixon, and W. Reuther. 1942. Some factors affecting rate of date leaf elongation. Proc. Amer. Soc. Hort. Sci. 41:71–84.

Aldrich, W. W., J. R. Furr, C. L. Crawford, and D. C. Moore. 1946. Checking of fruits of the Deglet Noor date in relation to water deficit in the palm. J. Agr. Res. 72:211–231.

Aldrich, W. W., and T. R. Young. 1941. Carbohydrate changes in the date palm during the summer. Proc. Amer. Soc. Hort. Sci. 39:110–118.

Anonymous. 1920. Les dattes de L'Algerie. Direcion de Territoires du Sud. Algerie Gouvt. Gen. Bull.

Aslan, S., R. Neja, D. Estrada, W. Dignon, and S. Mitchell. 1991. Soil, water, and climatic considerations in selecting date palm planting sites in the Coachella Valley. Indio, CA: Coachella Valley Resource Conservation District, Coachella Valley Water District, and Riverside County Farm Bureau.

Barger, W. R. 1933. Experiments with California dates in storage. Date Growers' Inst. Rep. 10:3–5.

Barnes, M. M., R. L. Warner, and E. F. Laird. 1984. Investigations on the biology and control of the carob moth on dates, Coachella Valley. Riverside, CA: UC Riverside Department of Entomology.

Bennett, J. F. 1950. Operations of marketing orders with particular reference to the marketing orders for dates. Date Growers' Inst. Rep. 27:31–32.

Berryman, E. 1972. 1971 Medjool date production in Bard, California. Date Growers' Inst. Rep. 49:10.

Bliss, D. E. 1944. *Omphalia* root rot of the date palm. Hilgardia 16:15–124.

Bliss, D. E., and R. O. Bream. 1940. Aeration as a factor in reducing fruit spoilage in dates. Date Growers' Inst. Rep. 17:11–15.

Bonavia, E. 1885. The future of the date palm in India. Calcutta: Thacker, Spink, and Co.

Boyden, B. L. 1930. Progress of *Parlatoria* date scale eradication. Date Growers' Inst. Rep. 7:16–19.

———. 1941. Eradication of *Parlatoria* date scale in the United States. Washington, DC: USDA Misc. Pub. 433.

Brooks, R. M., and H. P. Olmo. 1950. New fruit and nut varieties. Proc. Amer. Soc. Hort. Sci. 56:519.

Broschat, T. K. 1999. Nutrition and fertilization of palms. Palms 43:73–76.

Broschat, T. K., and A. W. Meerow. 2000. Ornamental palm horticulture. Gainesville FL: University Press of Florida.

Brown, G. K., R. M. Perkins, and E. G. Vis. 1969. An improved pesticide duster for date palms. Date Growers' Inst. Rep. 46:21–24.

Burkner, P. F., and R. M. Perkins. 1975. Mechanical extraction of date pollen. Date Growers' Inst. Rep. 52:3–7.

Byrd, N. P., R. E. Blair, and H. C. Phillips. 1947. California fruit and nut crop acreage estimates as of 1946. Calif. Dept. Agr. Bull. 36(2):2–30.

California Date Commission. 1998. A comprehensive systems approach for managing pests of California dates using low-risk strategies. Proposal to Pest Management Alliance, California Department of Pesticide Regulation. Indio, CA: California Date Commission.

Carpenter, J. B. 1979. Breeding date palms in California. Date Growers' Inst. Rep. 54:13–16.

Carpenter, J. B., and H. S. Elmer. 1978. Pests and diseases of the date palm. USDA Agric. Handbook 527.

Chase, A. R., and T. K. Broschat (eds.). 1991. Diseases and disorders of ornamental palms. St. Paul, MN: Amer. Phytopath. Soc.

Codekas, E. J., W. Geissler, D. H. Mitchell, and T. R. Brown. 1959. A panel discussion of current grower problems. Date Growers' Inst. Rep. 36:22–24.

Colley, C. C. 1967. The California date growing industry, 1890–1939. South. Calif. Quart. Parts 1 & 2. 49(1):47–63 and 49(2):167–191.

Cook, W. W. 1959. The future of the date industry in Coachella Valley. Date Growers' Inst. Rep. 36:20–22.

———. 1969. Looking at 10 years of date acreage changes. Date Growers' Inst. Rep. 46:4–5.

Davies, J., and P. A. Mauk. 1997. PMA pest management evaluation. Proposal to the California Date Commission. Indio, CA: California Date Commission.

Djerbi, M. 1982. Bayoud disease in North Africa: History, distribution, diagnosis and control (a review). Date Palm J. 1(2):153–198.

———. 1995. Précis de phoeniciculture. Rome: United Nations FAO.

———. 2003. Major diseases of the date palm. pp. 105–142 *in:* The date palm: From traditional resource to green wealth. Abu Dhabi: Emirates Center for Strategic Studies and Research.

Dowson, V. H. W. 1923. The varieties of date palms of the Shatt al'Arab. Dates and date cultivation of 'Iraq. Mesopotamia Dept. Agr. Mem. III, part III.

———. 1939. Provisional list of the dates of 'Iraq. Trop. Agr. 16:164–168.

———. 1982. Date production and protection. Plant Production and Protection Paper 35. Rome: United Nations FAO.

Elmer, H. S. 1964. Present status of date pest control studies. Date Growers' Inst. Rep. 41:4–6.

———. 1965. Banks grass mite, *Oligonychus pratensis,* on dates in California. J. Econ. Ent. 58:531–534.

———. 1966. Date palm insect and mite pests in the United States. Date Growers' Inst. Rep. 43:9–14.

Embleton, T. W., and J. A. Cook. 1947. The fertilizer value of date leaf and fruit stalk prunings. Date Growers' Inst. Rep. 24:18–19.

Fairchild, D. G. 1903. Persian Gulf dates and their introduction into America. Washington, DC: US Bur. Plant Indus. Bull. 54.

Farrar, K. 2005. Crop profiles for dates in California. (accessed August 17, 2005 at http://pestdata.ncsu.edu/cropprofiles/docs/cadates.html).

Fawcett, H. S., and L. J. Klotz. 1932. Diseases of the date palm, *Phoenix dactylifera.* Calif. Agr. Expt. Sta. Bull. 522.

Franklin, R. L. 1929. Present status of the date industry in Arizona. Date Growers' Inst. Rep. 6:7–8.

Furr, J. R., and W. W. Armstrong. 1955. Growth and yield of Khadrawy date palms irrigated at different intervals for two years. Date Growers' Inst. Rep. 32:3–7.

Furr, J. R., and H. D. Barber. 1950. The nitrogen content of some date garden soils in relation to soil management practices. Date Growers' Inst. Rep. 27:26–30.

Furr, J. R., E. C. Currlin, and W. W. Armstrong. 1952. Effects of water shortage during ripening and of nitrogen fertilization on yield and quality of Khadrawy dates. Date Growers' Inst. Rep. 29:10–12.

Furr, J. R., E. C. Currlin, R. H. Hilgeman, and W. Reuther. 1951. An irrigation and fertilization experiment with Deglet Noor dates. Date Growers' Inst. Rep. 28:17–20.

Furr, J. R., and A. A. Hewitt. 1964. Thinning trials on Medjool dates: Pollen dilution and chemical. Date Growers' Inst. Rep. 41:17–18.

Furr, J. R., and C. L. Ream. 1967. Growth and salt uptake of date seedlings in relation to salinity of the irrigation water. Date Growers' Inst. Rep. 44:2–4.

Geissler, W. 1971. Early history of the date industry in the United States. (unpublished manuscript).

Gerard, B. 1932. The effect of heat on the germination of date pollen. Date Growers' Inst. Rep. 9:15.

Glasner, B. 2004. Growing dates in Israel. Fruit Gard. 36(3):20–23.

Haas, A. R. C., and D. E. Bliss. 1935. Growth and composition of Deglet Noor dates in relation to water injury. Hilgardia 9:295–344.

Hansen, E. 2004. Looking for the Khalasah. Saudi Aramco World 55(4):2–8.

Hilgeman, R. H. 1972. History of date culture and research in Arizona. Date Growers' Inst. Rep. 49:11–14.

Hilgeman, R. H., and J. G. Smith. 1938. Maturation and storage studies with soft varieties of dates. Date Growers' Inst. Rep. 15:14–17.

Hodel, D. R. 1985. *Gliocladium* and *Fusarium* diseases of palms. Principes 29:85–88.

Hodel, D. R., and D. R. Pittenger. 2003a. Studies on the establishment of date palm (*Phoenix dactylifera* 'Deglet Noor') offshoots. Part I. Observations on root development and leaf growth. Palms 47:191–200.

———. 2003b. Studies on the establishment of date palm (*Phoenix dactylifera* 'Deglet Noor') offshoots. Part II. Size of offshoot. Palms 47:201–205.

Howard, F. W., D. Moore, R. M. Giblin-Davis, and R. G. Abad. 2001. Insects on palms. Wallingford UK: CABI Publishing.

Johnson, D. V., E. Joyal, and R. K. Harris. 2002. Date palm varieties in Arizona. Fruit Gard. 34(5):6–9, 26.

Karp, D. 2002. California dates: Ancient, delicious fruit of the palm. Fruit Gard. 34(5):14–18.

Kearney, T. H. 1906. Date varieties and date culture in Tunis. Washington DC: US Bur. Plant Indus. Bull. 92.

Kenknight, G. 1948. Findings of the *Omphalia* date root rot survey. Date Growers' Inst. Rep. 25:5–11.

Kenknight, G., and R. O. Amling. 1947. Progress report on the *Omphalia* date root rot survey. Date Growers' Inst. Rep. 24:10–17.

King, C. J., R. E. Beckett, and O. Parker. 1938. Agricultural investigations at the United States Field Station, Sacaton, Arizona, 1931–35. Washington DC: USDA Cir. 479.

Lindgren, D. L., D. E. Bliss, and D. F. Barnes. 1948. Insect infestation and fungus spoilage of dates: Their relation and control. Date Growers' Inst. Rep. 25:12–17.

Mahmud, T. H. 1958. Processing and marketing aspects of the date industry in the Coachella Valley, California. Thesis. University of California, Los Angeles.

Martius, C. F. P. von. 1823–1850. Historia naturalis palmarum. Munich, Germany.

Mason, S. C. 1915a. Botanical characters of the leaves of the date palm used in distinguishing cultivated varieties. Washington DC: USDA Bull. 223.

———. 1915b. Dates of Egypt and the Sudan. Washington DC: USDA Bull. 271.

———. 1927. Date culture in Egypt and the Sudan. Washington DC: USDA Bull. 1457.

Mauk, P. A., and S. Bellows. 1998. Report on control of Banks grass mite on date palms. Riverside CA: Univ. Calif., Coop. Ext., Riverside County.

McGeorge, W. T. 1954. Gypsum: A soil corrective and soil builder. Tempe, AZ: Ariz. Agr. Expt. Sta. Bull. 200.

———. 1955. Sulfur: A soil corrective and soil builder. Tempe AZ: Ariz. Agr. Expt. Sta. Bull. 201.

Millar, J. G., and H. H. Shorey. 1998. Mating disruption of carob moth in dates. Project Report for Calif. Dept. Pest. Regulation.

Mitchell, S. L. 1818. An encouragement to the introduction of the date-bearing palm into the United States. Amer. Monthly Mag. Critical Rev. 4:49–50.

Moore, D. C. 1938. The size of date fruit as affected by soil moisture. Date Growers' Inst. Rep. 15:3–4.

Nixon, R. W. 1928. The direct effect of pollen on the fruit of the date palm. J. Agr. Res. 36:97–128.

———. 1932. Observations on the occurrence of blacknose. Date Growers' Inst. Rep. 9:3–4.

———. 1933. Notes on rain damage to varieties at the US Experiment Date Garden. Date Growers' Inst. Rep. 10:10–13.

———. 1934. The Dairee date, a promising Mesopotamian variety for testing in the Southwest. Washington DC: USDA Cir. 300.

———. 1935. Metaxenia in dates. Proc. Amer. Soc. Hort. Sci. (1934) 32:221–226.

———. 1936. Metaxenia and interspecific pollinations in *Phoenix*. Proc. Amer. Soc. Hort. Sci. (1935) 33:21–26.

———. 1938a. Leaf pruning and fruit thinning following the freeze of January, 1937. Date Growers' Inst. Rep. 15:25–27.

———. 1938b. Discussion of the later effects of the freeze of January, 1937. Date Growers' Inst. Rep. 15:27–29.

———. 1940. Fruit thinning of dates in relation to size and quality. Date growers' Inst. Rep. 17:27–29.

———. 1942. Fruit shrivel of Halawy date in relation to amount and method of bunch thinning. Proc. Amer. Soc. Hort. Sci. 41:85–92.

———. 1943. Flower and fruit production of the date palm in relation to the retention of older leaves. Date Growers' Inst. Rep. 20:7–8.

———. 1946a. He brought African dates to Coachella. Desert Mag. 9:15–19.

———. 1946b. Bunch protection of the Khadrawy date in relation to sunburn and fruit shrivel. Date Growers' Inst. Rep. 23: 10–12.

———. 1947. Can a date palm carry too many leaves? Date Growers' Inst. Rep. 24:23–27.

———. 1950. Imported varieties of dates in the United States. Washington DC: USDA Cir. 834.

———. 1951. Fruit thinning experiments with the Medjool and Barhee varieties. Date Growers' Inst. Rep. 29:14–17.

———. 1955. American varieties of dates (unpublished manuscript).

———. 1956a. Effect of metaxenia and fruit thinning on size and checking of Deglet Noor dates. Proc. Amer. Soc. Hort. Sci. 67:258–264.

———. 1956b. How many fruits per strand should be left in thinning the Medjool date. Date Growers' Inst. Rep. 33:14.

———. 1957. Experimental planting of imported varieties of dates in the San Joaquin Valley of California. Date Growers' Inst. Rep. 34:15–16.

———. 1964. Rain damage to dates in 1963. Date Growers' Inst. Rep. 41:6–7.

———. 1971. Early history of the date industry in the United States. Date Growers' Inst. Rep. 48:26–30.

Nixon, R. W., and J. B. Carpenter. 1978. Growing dates in the United States. Washington DC: USDA Infor. Bull. 207.

Nixon, R. W., and C. L. Crawford. 1937. Fruit thinning experiments with Deglet Noor dates. Proc. Amer. Soc. Hort. Sci. 34:107–115.

———. 1942. Quality of Deglet Noor date fruits as influenced by bunch thinning. Proc. Amer. Soc. Hort. Sci. 34:103–110.

Nixon, R. W., and W. Reuther. 1947. The effect of environmental conditions prior to ripening on maturity and quality of date fruit. Proc. Amer. Soc. Hort. Sci. 49:81–91.

Nixon, R. W., and R. T. Wedding. 1956. Age of date leaves in relation to efficiency of photosynthesis. Proc. Amer. Soc. Hort. Sci. 67:265–269.

Oihabi, A. 2003. Major pests of the date palm. pp. 143–147 *in*: The date palm: From traditional resources to green wealth. Abu Dhabi: Emirates Center for Strategic Studies and Research.

Peightal, B. J. 1956. The date marketing order. Date Growers' Inst. Rep. 33:11–12.

Perkins, R. M., and P. F. Burkner. 1973. Mechanical pollination of date palms. Date Growers' Inst. Rep. 50:4–6.

Perring, T. M., C. Gispert, and C. A. Farrar. 2000. Development of IPM strategies for Banks grass mite (BGM) on dates in the Coachella Valley. Final Report to the California Date Commission.

Popenoe, P. B. 1912. Date growing in California. pp. 13–33 *in*: G. W. Jones, P. B. Popenoe, and R. D. Cornell, Date culture in Southern California. Los Angeles: Out West.

———. 1913a. Babylonian dates for California. Pomona Coll. J. Econ. Bot. 3:459–477.

———. 1913b. Date growing in the Old World and the New. Altadena, CA: West India Gardens.

———. 1973. The date palm. Miami, FL: Field Research Projects, Coconut Grove.

Postlethwaite, R. H. 1938. The Coachella Valley and its date industry. Coachella CA: V. V. Green.

Powers, H. B. 1945. Date production in Arizona. Tucson, AZ: U. Ariz. Agr. Ext. Serv. Cir. 125.

Ream, C. L. 1975. Date palm breeding—a progress report. Date Growers' Inst. Rep. 52:8–9.

Reuther, W. 1946. The effect of temperature and bagging on fruit set of dates. Date Growers' Inst. Rep. 23:3–7.

Ridgway, R. 1912. Color standards and color nomenclature. Washington DC: (self-published).

Rock, R. C. 1950. Market fruits under agreements and orders. Date Growers' Inst. Rep. 27:30–31.

Rygg, G. L. 1942. Factors affecting sugar spotting in dates. Date Growers' Inst. Rep. 19:10–12.

———. 1975. Date development, handling, and packing in the United States. Washington DC: USDA Agr. Handbook 482.

Shumway, N. P. 1960. Your desert and mine. Los Angeles: Westernlore Press.

Sievers, A. F., and W. R. Barger. 1930. Experiments on the processing and storing of Deglet Noor dates in California. Washington DC: USDA Tech. Bull. 193.

Simon, H. 1978. The date palm: Bread of the desert. New York: Dodd Mead & Co.

Stickney, F. S., D. F. Barnes, and P. Simmons. 1950. Date palm insects in the United States. Washington DC: USDA Dept. Agr. Cir. 846.

Swingle, L. 1950. Symposium on 1949–50 frost damage to date palms. Date Growers' Inst. Rep. 27:33–35.

Swingle, W. T. 1901. The date palm and its culture. USDA Yearbook of Agriculture 1900, pp. 453–490. Washington DC: USDA.

———. 1904. The date palm and its utilization in the southwestern US. Washington DC: Bur. Plant. Indus. Bull. 53.

———. 1945. Introduction of the Medjool date from Africa into the United States. Date Growers' Inst. Rep. 22:15–16.

Tate, H. F., and R. H. Hilgeman. 1971. Dates in Arizona. Tucson AZ: U. Ariz. Coop. Ext. Serv. Bull. A-22.

Thackery, F. A. 1952. A few notes on the Medjool date during its isolation in Nevada. Date Growers' Inst. Rep. 29:8–10.

Toumey, J. W. 1898. The date palm. Tempe AZ: Ariz. Agr. Exp. Sta. Bull. 29.

USDA. 1955. United States standards for grades of dates. Washington DC: Agricultural Marketing Service, USDA.

———. 1977. Grading manual: Dates and date products. Washington DC: Food Safety and Quality Service, USDA.

———. 2004. California Date Production. Washington DC: National Agricultural Statistics Service, USDA. Published on the Internet: http://www.nass.usda.gov/Statistics_by_State/California/Historical_Data/Dates.pdf (accessed December 15, 2004, 13:00 GMT).

Van Zyl, H. J. 1983. Date palm cultivation in South Africa. Fruit and Fruit Technology Research Institute, Department of Agriculture. Stellenbosch, Republic of South Africa.

Vincent, L. E., and D. L. Lindgren. 1958. Control of the date mite, *Oligonchus pratensis* (Banks), in California. Date Growers' Inst. Rep. 35:15–17.

———. 1972. The use of malathion for the control of date insects. Date Growers' Inst. Rep. 49:9.

Whittlesey, H. R. 1933. Ripening dates early by using different pollen. Date Growers' Inst. Rep. 10:9.

Wickson, E. J. 1889. The California fruits and how to grow them. San Francisco: Dewey & Co.

Winder, G. 1968. Production and marketing of Medjool dates in Bard, California. Date Growers' Inst. Rep. 45:10–11.

Wood, J. F., and E. Mortensen. 1938. Adaptability studies with date palms in southwest Texas. Proc. Amer. Soc. Hort. Sci. (1937) 35:231–234.

World Checklist of Monocots. 2005. The Board of Trustees of the Royal Botanic Gardens, Kew. Published on the Internet: http://www.kew.org/monocotChecklist/default.jsp (accessed 15 December 2005; 13:00 GMT).

Zaid, A. (ed.). 2002. Date palm cultivation. Rome: United Nations FAO Plant Production and Protection Paper 156, Rev. 1.

Zohary, D., and M. Hopf. 2000. Domestication of plants in the Old World. 3rd ed. Oxford: Oxford University Press.

Index